职业教育计算机网络技术专业
校企互动应用型系列教材

信息通信网络运行管理员（三级）实训教程

张文库　黄新颖　陈小明 ◎ 主　编
严宗浚　陈　铁　彭　锦 ◎ 副主编

电子工业出版社
Publishing House of Electronics Industry
北京·BEIJING

内 容 简 介

本书以《国家职业技能标准：信息通信网络运行管理员》为依据，紧紧围绕"以企业需求为导向、以职业能力为核心"的编写理念，突出职业培训特色；详细地介绍信息通信网络运行管理员（三级）应掌握的相关技能。全书分为 40 个任务，主要包括网络设备的安装与配置、服务器 Ubuntu 网络操作系统的安装与配置等。

本书结构合理，内容丰富，实用性强，可作为信息通信网络运行管理员（三级）的职业技能等级认定培训参考用书，也可供相关人员参加在职培训、岗位培训时使用。

未经许可，不得以任何方式复制或抄袭本书之部分或全部内容。
版权所有，侵权必究。

图书在版编目（CIP）数据

信息通信网络运行管理员（三级）实训教程 / 张文库，黄新颖，陈小明主编. -- 北京 : 电子工业出版社，2024. 12. -- ISBN 978-7-121-49334-8
Ⅰ．TN915
中国国家版本馆 CIP 数据核字第 2024F11H68 号

责任编辑：罗美娜
印　　刷：三河市华成印务有限公司
装　　订：三河市华成印务有限公司
出版发行：电子工业出版社
　　　　　北京市海淀区万寿路 173 信箱　　邮编：100036
开　　本：880×1230　　1/16　　印张：14.75　　字数：321 千字
版　　次：2024 年 12 月第 1 版
印　　次：2025 年 7 月第 2 次印刷
定　　价：45.80 元

凡所购买电子工业出版社图书有缺损问题，请向购买书店调换。若书店售缺，请与本社发行部联系，联系及邮购电话：（010）88254888，88258888。

质量投诉请发邮件至 zlts@phei.com.cn，盗版侵权举报请发邮件至 dbqq@phei.com.cn。
本书咨询联系方式：（010）88254617，luomn@phei.com.cn。

前 言

为适应新一代信息通信技术发展对技术人才、技能人才提出的新要求，进一步引领技能人才培养模式，夯实人才培养基础，本书的编写团队以《国家职业技能标准：信息通信网络运行管理员》为依据，紧紧围绕"以企业需求为导向、以职业能力为核心"的理念编写此书，从而满足职业技能培训与鉴定考核的需要。

1. 本书特色

本书按照《国家职业技能标准：信息通信网络运行管理员》规定的职业技能等级，编写与职业技能相关的知识内容，力求通俗易懂、深入浅出、灵活实用地让读者掌握信息通信网络运行管理员职业的主要技能。

本书的编写团队主要由企业一线的专业技术人员及长期从事职业能力水平评价工作的技工院校骨干教师组成，确保本书的内容能在职业技能及专业知识等方面得到最佳组合，并且能够突出技能人员培养与评价的特殊需求。

2. 教学资源

为了提高学生的学习效率和教师的教学效果，本书配备了电子课件和视频等教学资源。请有此需要的读者登录华信教育资源网免费注册后进行下载，若有问题，可在网站留言板留言，或者与电子工业出版社联系。

3. 本书编者

本书由珠海市技师学院张文库、广州市工贸技师学院黄新颖和广东省机械技师学院陈小明担任主编，广州市工贸技师学院严宗浚、东莞市技师学院陈铁和广东省技师学院彭锦担任副主编，参加编写的人员还有汕头技师学院曾楚生和清远市技师学院梁少峰。其中，陈铁编写了任务1至任务4，黄新颖编写了任务5至任务10，曾楚生编写了任务11至任务14，梁少峰编写了任务 15 至任务 18，严宗浚编写了任务 19 至任务 23，彭锦编写了任务 24 至 28，陈小明编写了任务 29 至任务 33，张文库编写了任务 34 至任务 40。全书由张文库、黄新颖和陈小明负责统稿及审校。

由于编者水平有限，加之时间仓促，书中难免存在疏漏之处，恳请广大读者批评指正。

编　者

目 录

网络设备部分

任务 1 部门之间的 IP 地址规划 2

任务 2 合理使用 IP 地址 6

任务 3 使用 eNSP 搭建和配置网络 11

任务 4 熟悉 VRP 基本操作 17

任务 5 交换机的基本配置 22

任务 6 实现不同部门的计算机之间的网络隔离 29

任务 7 实现相同部门的计算机跨交换机的网络互访 34

任务 8 利用三层交换机实现部门计算机之间的网络互访 40

任务 9 提高骨干链路的带宽 45

任务 10 实现网络负载均衡 48

任务 11 实现部门计算机动态获取 IP 地址 56

任务 12 提高网络的稳定性 61

任务 13 路由器的基本配置 68

任务 14 利用单臂路由实现部门计算机之间的网络互访 72

任务 15 使用静态路由实现网络连通 75

任务 16 使用默认及浮动路由实现网络连通 80

任务 17 使用动态路由 RIPv2 协议实现网络连通 85

任务 18 使用动态路由 OSPF 实现网络连通 89

任务 19	组建直连式二层无线局域网	93
任务 20	组建旁挂式三层无线局域网	103
任务 21	实现网络设备的远程管理	112
任务 22	使用基本 ACL 限制网络访问	117
任务 23	使用高级 ACL 限制服务器端口	121
任务 24	实现公司内网安全接入互联网	126
任务 25	利用静态 NAT 技术实现外网计算机访问内网服务器	131
任务 26	利用动态 NAPT 技术实现局域网计算机访问互联网	135

服 务 器 部 分

任务 27	安装与配置 Ubuntu 网络操作系统	140
任务 28	文件和目录管理	151
任务 29	软件包管理	158
任务 30	配置常规网络与 SSH 服务	162
任务 31	磁盘管理	169
任务 32	创建与管理软 RAID	177
任务 33	LVM 管理	181
任务 34	配置 DNS 服务器	187
任务 35	配置 DHCP 服务器	193
任务 36	配置 Apache 服务器	199
任务 37	配置 Nginx 服务器	205
任务 38	配置证书服务	210
任务 39	配置 FTP 服务器	216
任务 40	配置 MySQL 数据库服务器	222

网络设备部分

任务 1　部门之间的 IP 地址规划

任务目标

1．了解 IP 地址的表示和分类。
2．理解 IP 地址的分类。

任务描述

某公司内设有技术部、学术部、销售部这 3 个部门，每个部门有 20 台计算机，ISP 已分配地址段 192.168.10.0/24 给该公司使用，请充分考虑网络的性能及管理效率等因素，对该网络的 IP 地址进行规划。

任务要求

由任务描述可知，公司的 3 个部门拥有计算机数量均为 20 台，且从 ISP 处获得 1 个 C 类 IP 地址段。从网络性能的方面考虑，应尽量缩减网络流量，将部门内部通信业务尽量"圈定"在部门内部；从日常管理的角度考虑，把 1 个较大的网络分成相对较小的网络有利于隔离和排除故障。因此，可以考虑通过合理的子网划分来解决这一问题。

该公司的网络拓扑结构如图 1.1 所示。

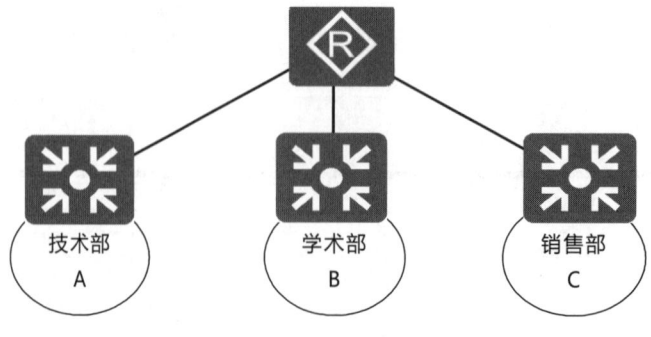

图 1.1　公司的网络拓扑结构

具体要求如下。

（1）确定各子网的主机数量：每台 TCP/IP 主机至少需要 1 个 IP 地址，路由器的每个端口都需要 1 个 IP 地址。

（2）确定每个子网的大小。

（3）根据任务需要，创建以下内容：为整个网络设定 1 个子网掩码；为每个物理网段设定 1 个子网 ID；为每个子网确定主机的合法 IP 地址范围。

任务实施

步骤 1：确定各子网的主机数量。IPv4 中的地址是由 32 位二进制位组成的，分为网络位和主机位两部分，IP 地址结构如图 1.2 所示。

图 1.2　IP 地址结构

由前面的分析可知，每个部门都需要 21 个 IP 地址，其中 20 个 IP 地址供计算机使用，1 个 IP 地址供路由器端口使用。

步骤 2：确定每个子网的大小。十进制数"21"至少需要 5 位二进制位来表达，于是可以确定子网的大小为 2^5。子网的大小示意如图 1.3 所示。

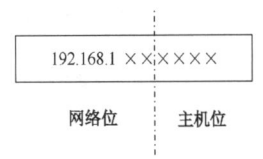

图 1.3　子网的大小示意

步骤 3：创建子网掩码、子网 ID、合法 IP 地址范围。

（1）为整个网络设定子网掩码。将图 1.3 中网络位的二进制值全部设置为"1"，主机位的二进制值全部设置为"0"，即可得到划分子网后的子网掩码，计算过程如图 1.4 所示。

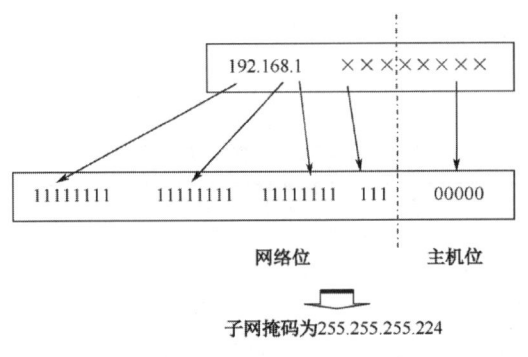

图 1.4　子网掩码的计算过程

（2）为每个物理网段设定 1 个子网 ID。RFC 标准规定，子网的网络 ID 不能全为"0"或全为"1"，合法的子网 ID 的计算过程如图 1.5 所示。

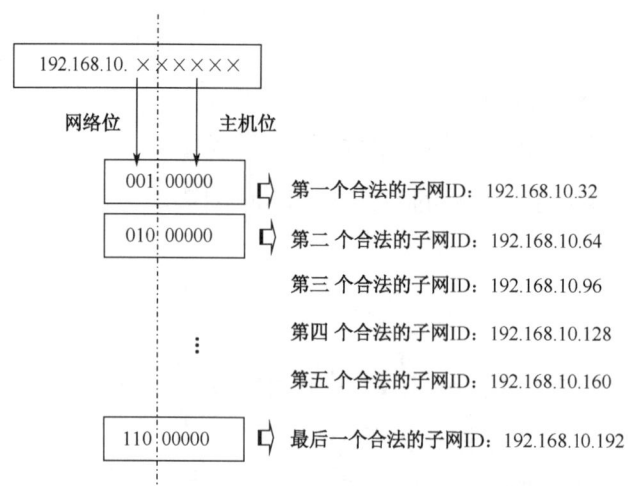

图 1.5　合法的子网 ID 的计算过程

（3）为每个子网确定主机的合法 IP 地址范围。RFC 标准规定，主机 ID 不能全为"0"或全为"1"，下面以第一个合法的子网为例说明子网中合法的主机 ID 的计算过程，如图 1.6 所示。

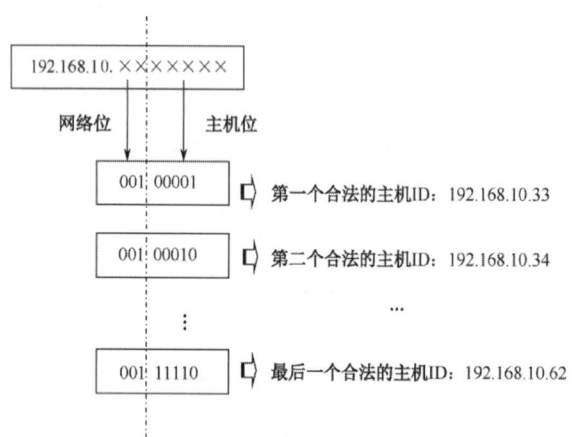

图 1.6　第一个合法的子网中合法的主机 ID 的计算过程

任务验收

（1）经计算得出本任务三个部门使用的子网中合法的主机 ID，如表 1.1 至表 1.3 所示。

表 1.1　子网 192.168.10.32

子　网	主　机	意　义
192.168.10.32	192.168.10.32	子网的网络地址
	192.168.10.33	子网中第一个合法的主机 ID
	192.168.10.62	子网中最后一个合法的主机 ID
	192.168.10.63	子网的广播地址

表 1.2 子网 192.168.10.64

子　　网	主　　机	意　　义
192.168.10.64	192.168.10.64	子网的网络地址
	192.168.10.65	子网中第一个合法的主机 ID
	192.168.10.94	子网中最后一个合法的主机 ID
	192.168.10.95	子网的广播地址

表 1.3 子网 192.168.10.96

子　　网	主　　机	意　　义
192.168.10.96	192.168.10.96	子网的网络地址
	192.168.10.97	子网中第一个合法的主机 ID
	192.168.10.126	子网中最后一个合法的主机 ID
	192.168.10.127	子网的广播地址

（2）经过以上计算得到该公司网络各部门的 IP 地址规划，如图 1.7 所示。

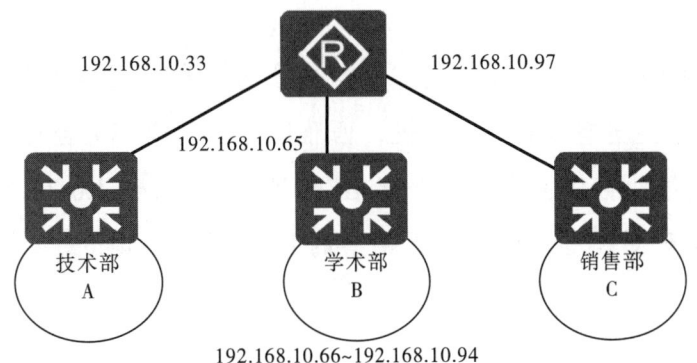

图 1.7 公司网络各部门的 IP 地址规划

任务 2　合理使用 IP 地址

任务目标

1．掌握 IP 地址的子网划分。
2．合理规划和使用 IP 地址。

任务描述

某跨国公司下设有"珠海总公司"、"广州分公司"和"西雅图分公司"。珠海总公司有 80 台计算机，广州分公司有 23 台计算机，西雅图分公司有 50 台计算机，且 ISP（Internet Service Provider，网络业务提供商）已分配 IP 地址段 192.168.1.0/24 给该公司使用。请充分考虑网络的性能及管理效率等因素，对该网络的 IP 地址进行规划。

任务要求

由任务描述可知，该公司的三个办公地点的计算机数量差异较大，珠海总公司所需的主机数量最多，至少应该划分一个大小为 96 个 IP 地址的地址块供其使用，如果依据划分子网的方法（定长子网），则所需的 IP 地址的数量为 96×3=288，但 ISP 只提供了一个 C 类 IP 地址段，IP 地址的数量为 255。由 255<288 可知，定长子网划分在本任务中无法实现，可以考虑通过变长子网划分来解决问题。

该公司的网络拓扑结构如图 2.1 所示。

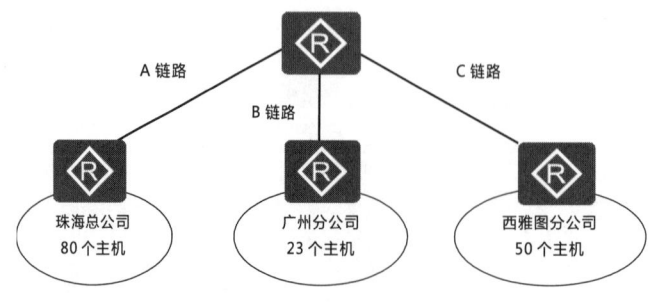

图 2.1　公司的网络拓扑结构

具体要求如下。

（1）确定各子网的主机数量：每台 TCP/IP 主机至少需要 1 个 IP 地址，路由器的每个端口都需要 1 个 IP 地址。

（2）确定每个子网的大小。

（3）根据任务需要，创建以下内容：为每个子网设定 1 个子网掩码；为每个物理网段设定 1 个子网 ID；为每个子网确定主机的合法 IP 地址范围。

步骤 1：确定各子网的主机数量，如表 2.1 所示。

表 2.1 主机数量

子 网	主 机 数 量
珠海总公司	80
广州分公司	23
西雅图分公司	50
A 链路	2
B 链路	2
C 链路	2

步骤 2：确定每个子网的大小，如表 2.2 所示。

表 2.2 子网大小

子 网	主 机 数 量	子 网 大 小	备 注
珠海总公司	80	128	—
广州分公司	23	32	—
西雅图分公司	50	64	—
A 链路	2	4	子网中至少需要 4 台主机 ID，否则除网络 ID 和广播地址外无 IP 地址可用
B 链路	2	4	
C 链路	2	4	

步骤 3：创建子网掩码、子网 ID、合法 IP 地址范围。

划分子网的思路如下：首先为较大子网分配地址块，然后在从未被分配的地址块中为剩下的较大子网分配地址块，以此类推（注意：此处将使用 1 个子网和 0 个子网），变长子网划分的思路如图 2.2 所示。

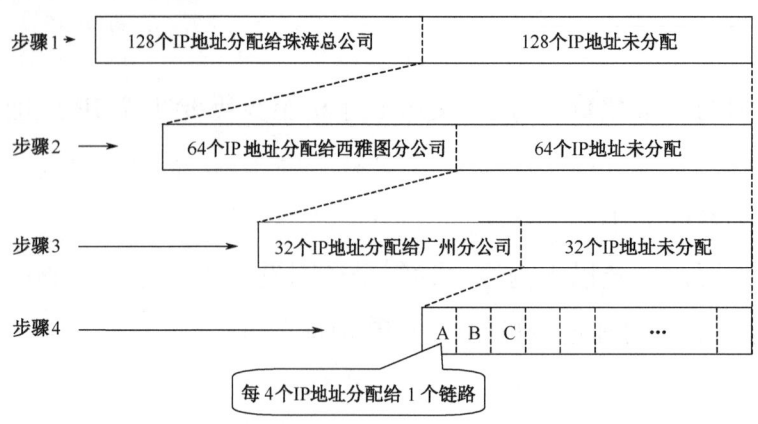

图 2.2 变长子网划分的思路

（1）为每个子网设定子网掩码。

珠海总公司所需地址块为 128，即需要 7 位二进制位，因此子网位为 1 位二进制位，子网掩码的计算过程如图 2.3 所示。

西雅图分公司所需地址块为 64，即需要 6 位二进制位，因此子网位为 2 位二进制位，子网掩码的计算过程如图 2.4 所示。

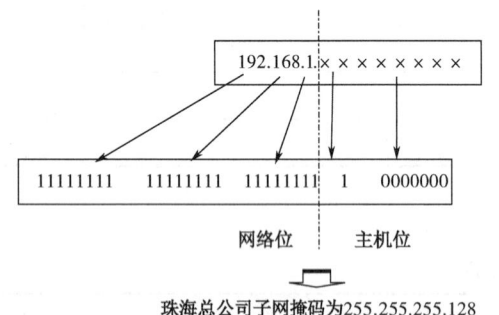

图 2.3 珠海总公司子网掩码的计算过程

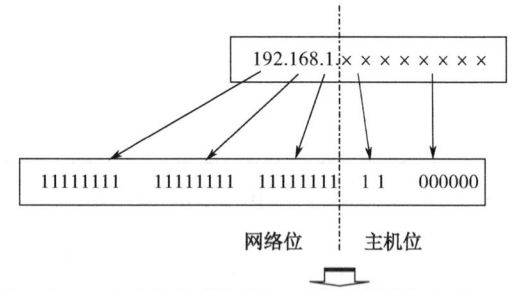

图 2.4 西雅图分公司子网掩码的计算过程

广州分公司所需地址块为 32，即需要 5 位二进制位，因此子网位为 3 位二进制位，子网掩码的计算过程如图 2.5 所示。

A、B、C 链路所需地址块为 4，即需要 2 位二进制位，因此子网位为 6 位二进制位，子网掩码的计算过程如图 2.6 所示。

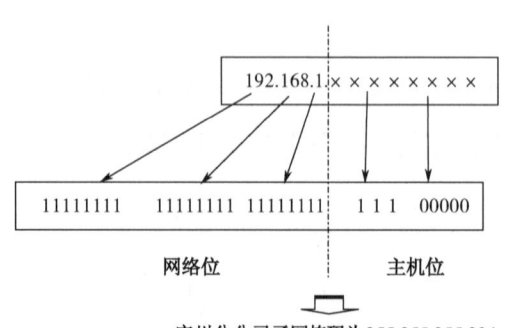

图 2.5 广州分公司子网掩码的计算过程

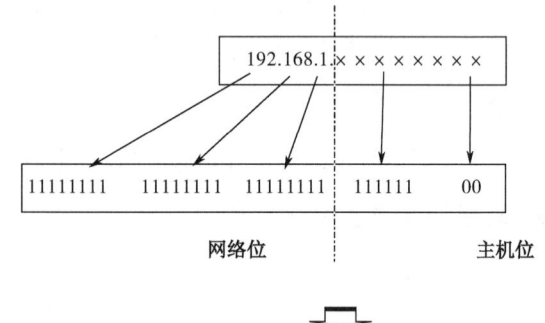

图 2.6 A、B、C 链路子网掩码的计算过程

（2）为每个物理网段设定 1 个子网 ID，计算过程如图 2.7 所示。

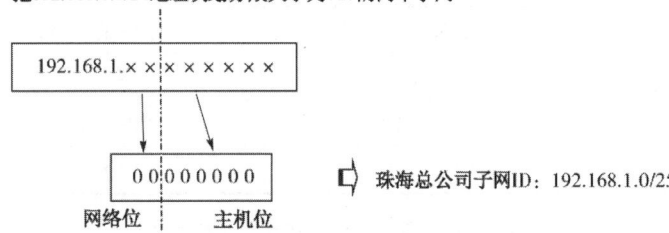
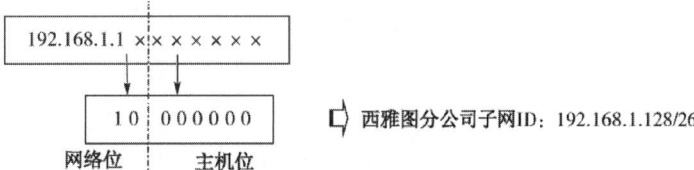
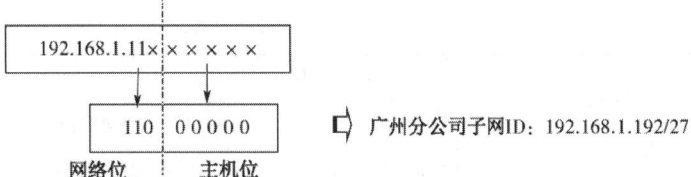
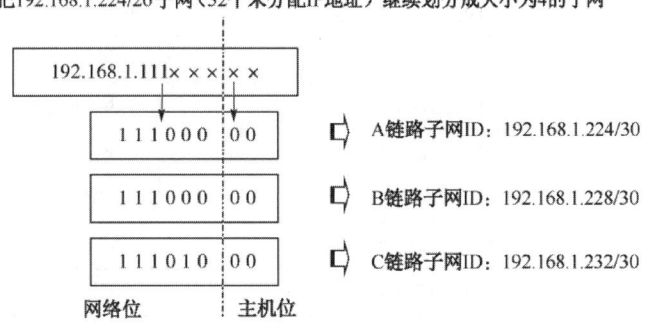

图 2.7　子网 ID 的计算过程

（3）为每个子网确定主机的合法 IP 地址范围。

任务验收

（1）经计算得出本任务 6 个子网的合法主机 ID，如表 2.3 至表 2.6 所示。

表 2.3　子网 192.168.1.0/25

子网	部门	主机	意义
192.168.1.0/25	珠海总公司	192.168.1.0/25	子网的网络地址
		192.168.1.1/25	子网中第一个合法的主机 ID
		192.168.1.126/25	子网中最后一个合法的主机 ID
		192.168.1.127/25	子网的广播地址

表 2.4　子网 192.168.1.128/26

子　网	部　门	主　机	意　义
192.168.1.128/26	西雅图分公司	192.168.1.128/26	子网的网络地址
		192.168.1.129/26	子网中第一个合法的主机 ID
		192.168.1.190/26	子网中最后一个合法的主机 ID
		192.168.1.191/26	子网的广播地址

表 2.5　子网 192.168.1.192/27

子　网	部　门	主　机	意　义
192.168.1.192/27	广州分公司	192.168.1.192/27	子网的网络地址
		192.168.1.193/27	子网中第一个合法的主机 ID
		192.168.1.222/27	子网中最后一个合法的主机 ID
		192.168.1.223/27	子网的广播地址

表 2.6　链路子网 A～C

子　网	部　门	主　机	意　义
192.168.1.224/30	A 链路	192.168.1.224/30	子网的网络地址
		192.168.1.225/30	子网中第一个合法的主机 ID
		192.168.1.226/30	子网中最后一个合法的主机 ID
		192.168.1.227/30	子网的广播地址
192.168.1.228/30	B 链路	192.168.1.228/30	子网的网络地址
		192.168.1.229/30	子网中第一个合法的主机 ID
		192.168.1.230/30	子网中最后一个合法的主机 ID
		192.168.1.231/30	子网的广播地址
192.168.1.232/30	C 链路	192.168.1.232/30	子网的网络地址
		192.168.1.233/30	子网中第一个合法的主机 ID
		192.168.1.234/30	子网中最后一个合法的主机 ID
		192.168.1.235/30	子网的广播地址

（2）经过以上计算得到该公司网络各部门的 IP 地址规划，如图 2.8 所示。

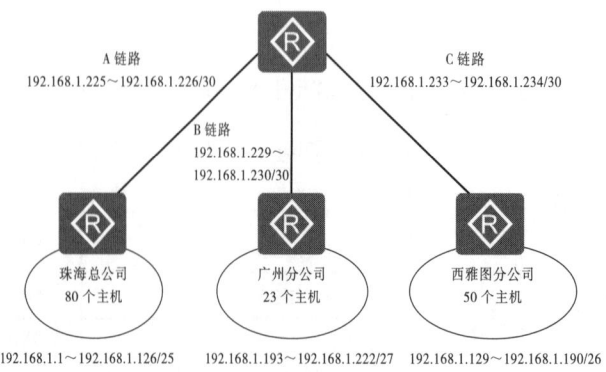

图 2.8　公司网络各部门的 IP 地址规划

任务 3 使用 eNSP 搭建和配置网络

任务目标

1. 认识 eNSP 主界面和网络连接的线缆。
2. 使用 eNSP 搭建和配置网络。

任务描述

在 eNSP 中进行一组网络实验，首先要搭建用于实验的网络拓扑结构，这就要求用户掌握如何在 eNSP 中添加网络设备，以及如何与相邻的网络设备进行连接。在 eNSP 模拟器中，可以利用图形化来灵活地搭建需要的拓扑图。

任务要求

在本任务中，读者应重点学习网络设备的添加与连线，基于路由器和交换机的网络拓扑结构如图 3.1 所示[①]。

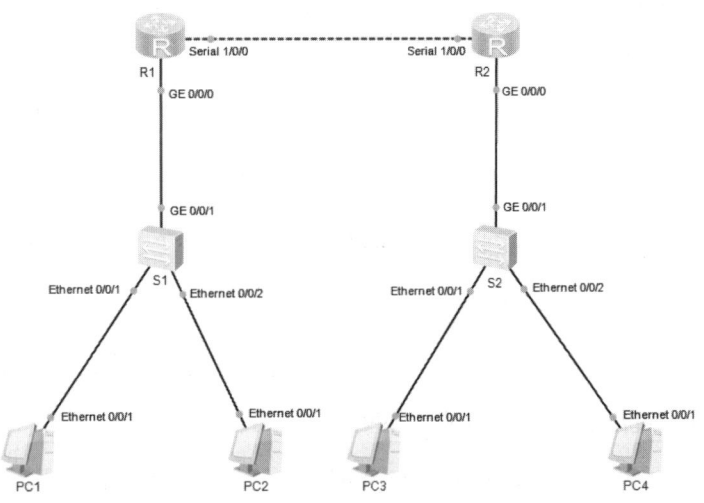

图 3.1 基于路由器和交换机的网络拓扑结构

---

① 在本书中，GE 表示 GigabitEthernet。

任务实施

步骤 1：添加网络设备。

eNSP 模拟器的主界面左侧为可供选择的网络设备区，从左至右、从上到下依次为路由器、交换机、无线局域网、防火墙、终端、其他设备和设备线缆，网络设备区如图 3.2 所示。

本任务需要添加一个型号为 AR2220 的路由器，将需要的设备直接拖曳至工作区。添加完成后在工作区可以看到一个标签名称为"AR1"的路由器图标。使用同样的方法可以添加其他网络设备，添加完成后可以通过鼠标拖曳的方式来调整各个网络设备之间的位置关系，工作区如图 3.3 所示。

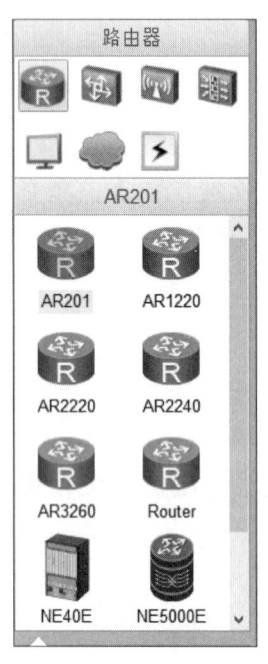

图 3.2 网络设备区（部分）

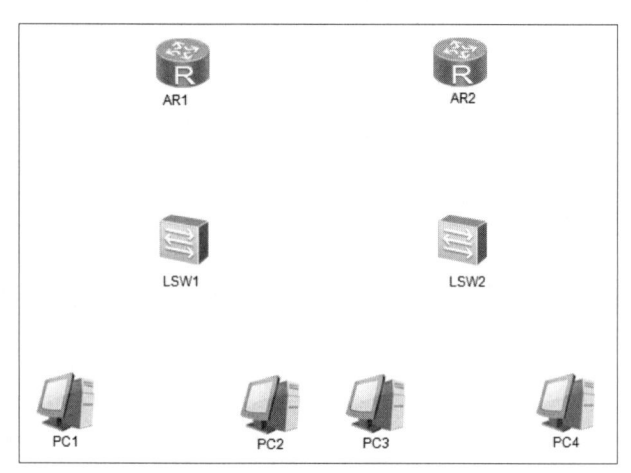

图 3.3 工作区

步骤 2：配置网络设备。

右击工作区中的网络设备，在弹出的快捷菜单中选择"设置"命令，打开网络设备配置界面。

（1）在"视图"选项卡中，可以查看网络设备面板及可供选择的 eNSP 支持的接口卡。如果需要为网络设备添加接口卡，则在"eNSP 支持的接口卡"选区选择合适的接口卡，并将其直接拖曳至上方的网络设备面板中相应的槽位即可。如果需要删除某个接口卡，则直接将网络设备面板中的接口卡拖曳回"eNSP 支持的接口卡"选区即可。注意，只有在网络设备电源关闭的情况下才能进行添加或删除接口卡的操作。

在本任务中添加"2SA"串口模块，网络设备配置界面如图 3.4 示。

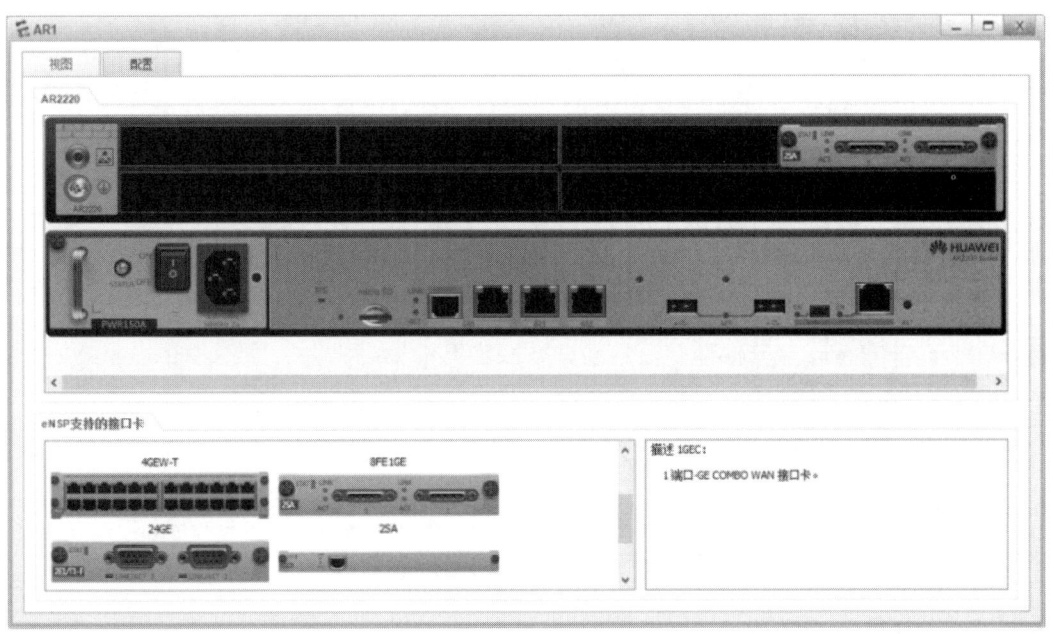

图 3.4 网络设备配置界面

(2)在"配置"选项卡中,可以配置网络设备的串口号,串口号范围为 2000~65 535,默认情况下串口号为 2000。更改串口号并单击"应用"按钮即可使新配置的串口号生效,如图 3.5 所示。

图 3.5 配置网络设备的串口号

(3)右击 PC1,在弹出的快捷菜单中选择"设置"命令,打开设置对话框。在"基础配置"选项卡中配置 PC1 的基础参数,如 IP 地址、子网掩码和 MAC 地址等,如图 3.6 所示。

图 3.6 配置 PC1 的基础参数

步骤 3：使用线缆连接设备。

当网络设备添加完成后，选择相应的线缆，并在要连线的网络设备上单击即可完成连接。在本任务中，AR1 与 AR2 连接时使用串口线。当单击 AR1 时会弹出如图 3.7 所示的端口选择界面，先选中要连接的端口，再将其拖曳至 AR2 上，选中适当的端口即可完成连接操作。

使用同样的方法，可以对其他网络设备之间的连接进行设置。本任务中所有网络设备之间都是使用直通线连接的。网络设备连线区如图 3.8 所示。

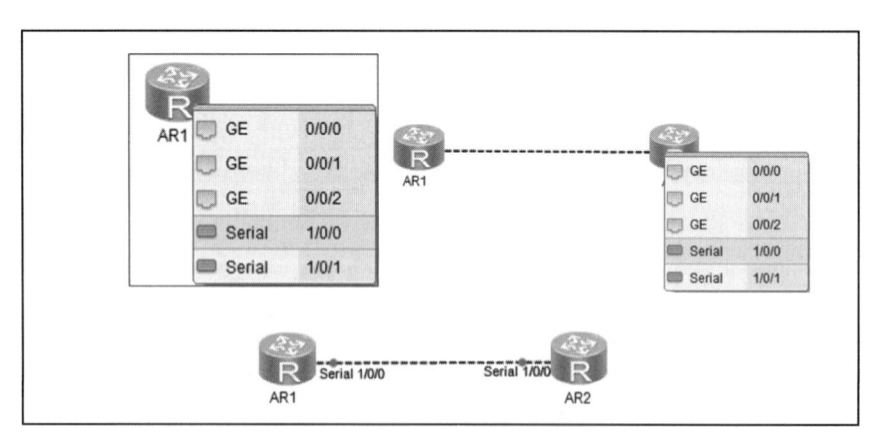

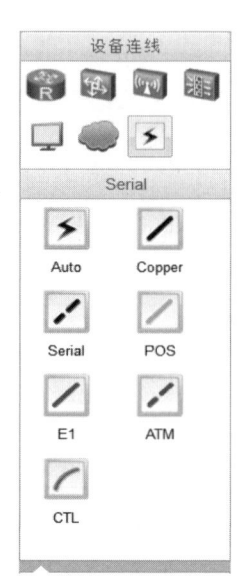

图 3.7 端口选择界面　　　　　　　　　　图 3.8 网络设备连线区

知识链接：

- 网络设备连接线缆有以下几种。

- Auto：自动识别接口卡选择相应的线缆。
- Copper：双绞线，连接网络设备的以太网接口。
- Serial：串口线，连接网络设备的串口。
- POS：POS 口连接线，连接路由器的 POS 口。
- E1：E1 口连接线，连接路由器的 E1 口。
- ATM：ATM 口连接线，连接路由器的 4G.SHDSL 口。

步骤 4：修改网络设备的标签名称。

单击网络设备的标签，可以进入标签的编辑状态，将 AR1 的标签名称改为"R1"，如图 3.9 所示。使用同样的方法，可以更改其他网络设备的标签名称，完成后的网络拓扑图应该与图 3.1 相似。

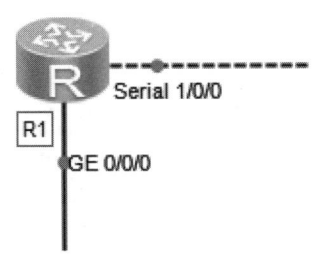

图 3.9　修改网络设备的标签名称

步骤 5：启动网络设备。

选中需要启动的网络设备后，可以通过单击工具栏中的"启动设备"按钮，或者右击该网络设备，在弹出的快捷菜单中选择"启动"命令来启动所需的网络设备，如图 3.10 所示；也可以全选所有网络设备并右击，在弹出的快捷菜单中选择"启动"命令，启动所有网络设备。在网络设备启动后，可以双击网络设备的图标，通过弹出的命令行界面对网络设备进行配置。

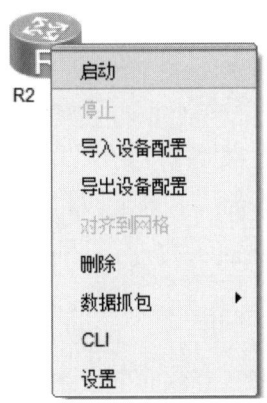

图 3.10　启动所需的网络设备

任务验收

完成网络搭建后，需要检查使用的线缆是否正确，是否连接了正确的端口，如图 3.11 所示。

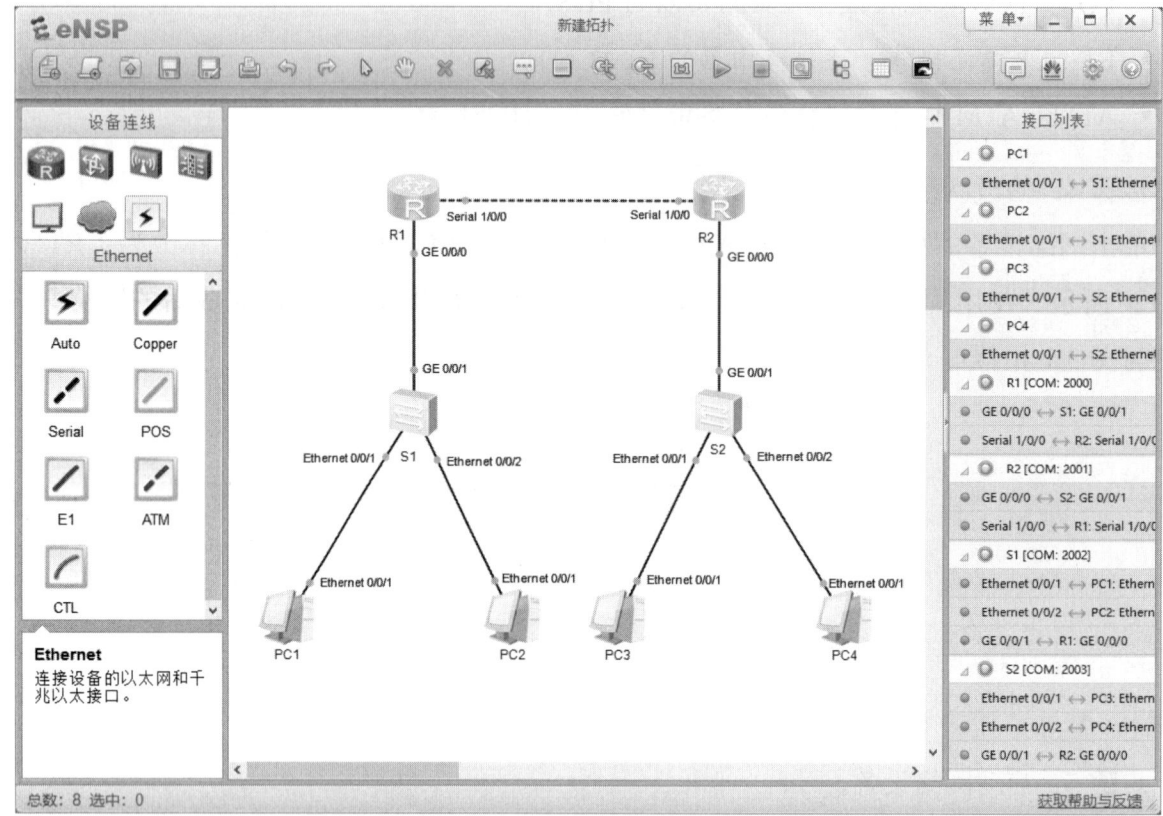

图 3.11　网络搭建完成

任务 4　熟悉 VRP 基本操作

任务目标

1. 熟悉 VRP 的命令视图。
2. 熟悉 VRP 基本操作。

任务描述

在用户登录交换机或路由器设备并出现命令行提示符后，即可进入 CLI（Command Line Interface，命令行界面）。命令行界面是用户与路由器进行交互的常用工具。

用户在输入命令时，如果不记得此命令的关键字或参数，则可以使用命令视图的在线帮助功能获取全部或部分关键字和参数的提示。用户也可以使用快捷键完成对应命令的输入，简化操作。

任务要求

本任务模拟用户首次使用 VRP（Versatile Routing Platform，通用路由平台）操作系统的过程。在登录路由器或交换机后，使用命令行界面配置网络设备，并使用不同视图的切换、命令视图的在线帮助功能和快捷键，完成网络设备的基本配置。

任务实施

步骤1：进入和退出命令视图。

在启动网络设备后（所选网络设备为路由器），双击该网络设备可以成功登录设备，进入用户视图，此时视图中的提示符是"\<Huawei\>"。

"quit"命令用于从任何视图退出到上一层视图。例如，端口视图是从系统视图进入的，所以系统视图是端口视图的上一层视图。

```
<Huawei>                              //用户视图
<Huawei>system-view                   //进入系统视图
```

```
Enter system view, return user view with Ctrl+Z.
[Huawei]interface GigabitEthernet 0/0/0         //进入端口视图
[Huawei-GigabitEthernet1/0/0]quit
[Huawei]                                        //已退出到系统视图
[Huawei]quit
<Huawei>                                        //已退出到用户视图
```

有些命令视图的层级很深，从当前视图退出到用户视图，需要多次执行"quit"命令。使用"return"命令（或者按"Ctrl+Z"快捷键），可以直接从当前视图退出到用户视图。

```
[Huawei-GigabitEthernet1/0/0]return
<Huawei>                                        //已退出到用户视图
```

知识链接：

VRP 的命令视图如图 4.1 所示。

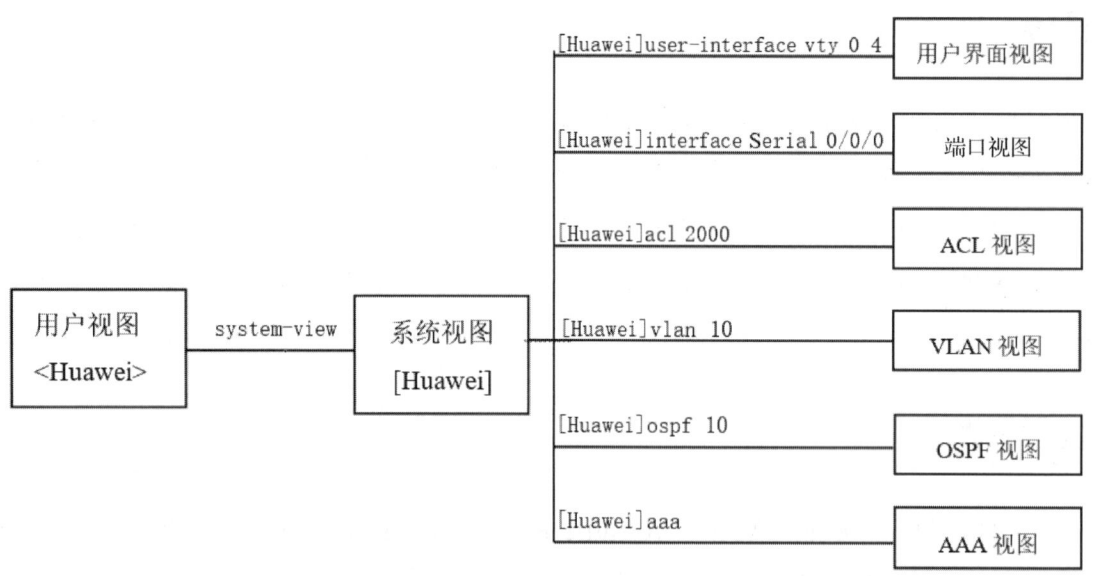

图 4.1　VRP 的命令视图

步骤 2：设置设备名称。

用户可以根据需要来使用命令提示符，在系统视图中使用"sysname"命令设置设备名称。

```
[Huawei]sysname R1                              //设置设备名称为"R1"
[R1]
```

步骤 3：设置系统时钟。

为了保证网络与其他网络设备协调工作，需要准确设置系统时钟。可以使用"clock timezone"命令设置所在时区，使用"clock datetime"命令设置当前系统的时间和日期。

```
<R1>clock timezone BJ add 08:00:00              //所在时区为北京
或
<R1>clock timezone CST add 08:00:00             //时区取名为 CST（中国标准时间）北京时间（UTC+8）
```

```
<R1>clock datetime 12:00:00 2023-11-11      //设置系统的时间为12时，日期为2023年11月11日
<R1>display clock
2023-11-11  12:00:05
Sunday
Time Zone(CST) : UTC+08:00
```

步骤4：设置标题信息。

```
[R1]header login information "Hello"
[R1]header shell information "Welcome to Huawei"
```

步骤5：查看网络设备的基本信息。

在用户视图中，只能使用参观和监控级的命令。例如，使用"display version"命令显示 VRP 操作系统的版本、设备的具体型号和启动时间等信息。

```
<R1>display version
Huawei Versatile Routing Platform Software
VRP (R) software, Version 5.130 (AR2200 V200R003C00)
Copyright (C) 2011-2012 HUAWEI TECH CO., LTD
Huawei AR2220 Router uptime is 0 week, 0 day, 0 hour, 6 minutes
BKP 0 version information:
1. PCB       Version   : AR01BAK2A VER.NC
2. If Supporting PoE : No
3. Board    Type      : AR2220
4. MPU Slot Quantity : 1
5. LPU Slot Quantity : 6
MPU 0(Master) : uptime is 0 week, 0 day, 0 hour, 6 minutes
MPU version information :
1. PCB       Version   : AR01SRU2A VER.A
2. MAB       Version   : 0
3. Board    Type      : AR2220
4. BootROM Version   : 0
//从上面的内容中可以看到VRP操作系统的版本、设备的具体型号和启动时间等信息
```

步骤6：配置网络设备的端口信息。

在端口视图中，可以使用"ip address"命令配置该端口的 IP 地址、子网掩码等。配置路由器 R1 的 GE 0/0/0 端口 IP 地址为 192.168.1.1，子网掩码为 24 位。

```
[R1]int GigabitEthernet 0/0/0
[R1-GigabitEthernet0/0/0]ip address 192.168.1.1 24
```

步骤7：使用"undo"命令。

（1）使用"undo"命令恢复默认配置。

```
[Huawei]sysname R1                           //设置设备名称为"R1"
[R1]undo sysname                             //恢复设备名称为"Huawei"
[Huawei]
```

（2）使用"undo"命令禁用某个功能。

[Huawei]undo stp enable   //禁用 STP

（3）使用"undo"命令删除某项配置。

[Huawei]interface GigabitEthernet 0/0/0   //进入端口视图
[Huawei-GigabitEthernet0/0/0]ip address 192.168.1.254 24   //配置端口 IP 地址
[Huawei-GigabitEthernet0/0/0]undo ip address   //删除端口 IP 地址

步骤 8：使用命令视图的在线帮助功能功能。

命令视图提供如下 2 种在线帮助功能。

（1）完全帮助。在任意一个命令视图中，输入"?"即可获取该命令视图中的所有命令及其简单描述。

```
<Huawei> ?
```

输入一个命令，并在后面输入空格和"?"，如果"?"的位置为关键字，则列出全部关键字及其简单描述。例如：

```
<Huawei> language-mode?
Chinese  Chinese environment
English  English environment
```

其中，Chinese、English 是关键字，Chinese environment 和 English environment 是对关键字的简单描述。

输入一个命令，并在后面输入空格和"?"，如果"?"的位置为参数，则列出有关参数的参数名和参数描述。例如：

```
[Huaweildisplay aaa ?
configuration  AAA configuration
[Huawei]display aaa configuration ?
<cr>
```

其中，configuration 是参数名，AAA configuration 是对参数的简单描述；而<cr>表示该位置无关键字或参数。

（2）部分帮助。输入一个字符串，并在后面输入"?"，则列出以该字符串开头的所有命令。

```
<Huawei> d?
debugging  delete  dir  display
```

输入一个命令，并在后面输入一个字符串和"?"，则列出以该字符串开头的所有关键字。

```
<Huawei> display v?
version  virtual-access  vlan  vpls  vrrp  vsi
```

输入命令某关键字的前几个英文字母，按"Tab"键，则列出完整的关键字。前提是在这个命令中，这几个英文字母仅与该关键字相关，不会与其他关键字混淆。

可以通过在用户视图中执行"language-mode Chinese"命令将以上帮助信息切换为中文。

任务验收

（1）先退出网络设备的用户视图，再重新进入用户视图，查看标题信息。

（2）使用"display current-configuration"命令查看当前配置，检验主机名称、时间设置和标题信息等是否正确。

（3）使用"display interface GigabitEthernet 0/0/0"命令查看端口信息。

（4）使用"display ip interface brief"命令查看端口 IP 地址信息。

任务 5　交换机的基本配置

任务目标

1．熟悉交换机的各种配置模式。
2．使用终端软件正确连接交换机。
3．熟练配置交换机的各项网络参数及端口状态。

任务描述

某公司因业务发展需求，购买了一批华为交换机来扩展现有的网络。按照公司网络管理的要求，网络管理员小赵需要通过交换机的 Console 口进行连接，完成交换机的配置和管理任务，并优化网络环境。

任务要求

本次任务包括几种视图模式的进入与退出、配置 Console 口密码、命名交换机、配置交换机恢复出厂状态、使用"?"查看帮助命令、配置日期时钟等。

（1）为了实现交换机的基本配置，网络拓扑结构如图 5.1 所示。

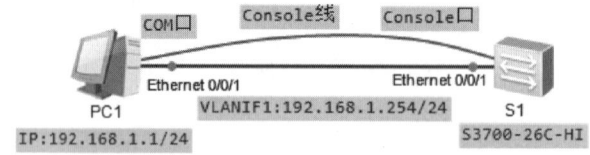

图 5.1　交换机基本配置的网络拓扑结构

（2）S1 和 PC1 的端口及 IP 地址设置如表 5.1 所示。

表 5.1　S1 和 PC1 的端口及 IP 地址设置

设备名称	本端端口	IP 地址/子网掩码	网　　关	对端设备与端口
S1	Ethernet 0/0/1	192.168.1.254/24	—	PC1:Ethernet 0/0/1
PC1	Ethernet 0/0/1	192.168.1.1/24	192.168.1.254	S1:Ethernet 0/0/1

（3）使用 CLI 管理交换机设备，对交换机进行基本配置。

任务实施

1. 交换机的管理方式配置

步骤 1：按照图 5.1 搭建网络拓扑，连线全部使用直通线，并开启所有设备电源。

步骤 2：双击 PC1，在打开的设置对话框中选择"串口"选项卡，如图 5.2 所示。

步骤 3：默认的超级终端参数设置如图 5.3 所示，单击"连接"按钮。

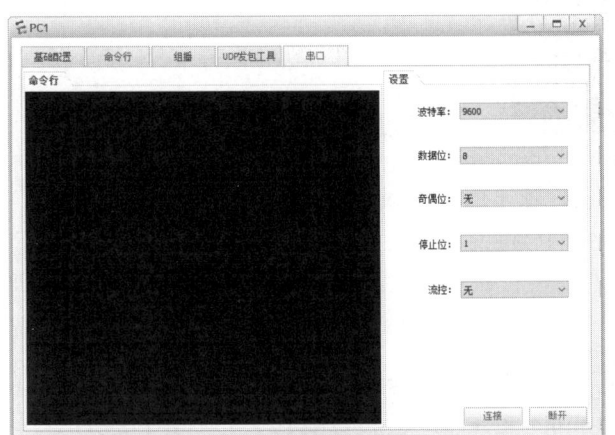
图 5.2 选择"串口"选项卡

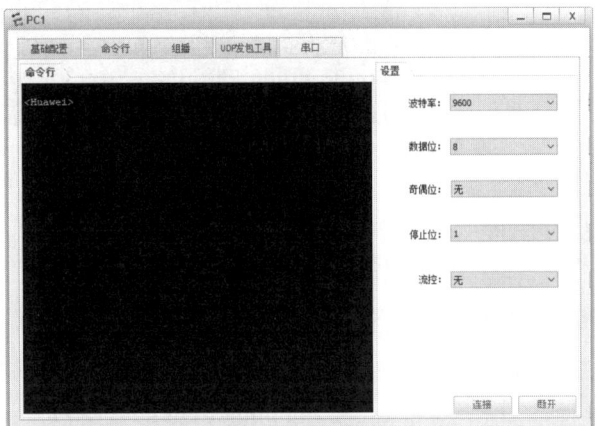
图 5.3 默认的超级终端参数设置

知识链接：

Console 口登录管理

在交换机上都有一个 Console 口，它是专门用于配置和管理交换机的端口。计算机通过 Console 口与交换机连接。Console 口的类型多为 RJ-45。通过专门的 Console 线将交换机的 Console 口连接到计算机的串口上，计算机可以成为超级终端。

设置 Console 口的波特率为"9600"，数据位为"8"，奇偶校验为"无"，停止位为"1"，流控为"无"。

步骤 4：此时，用户已经成功进入交换机的配置界面，在该界面中可以对交换机进行必要的配置。使用"display version"命令可以查看交换机的软/硬件版本信息，并使用"save"命令可以保存配置信息，如图 5.4 所示。

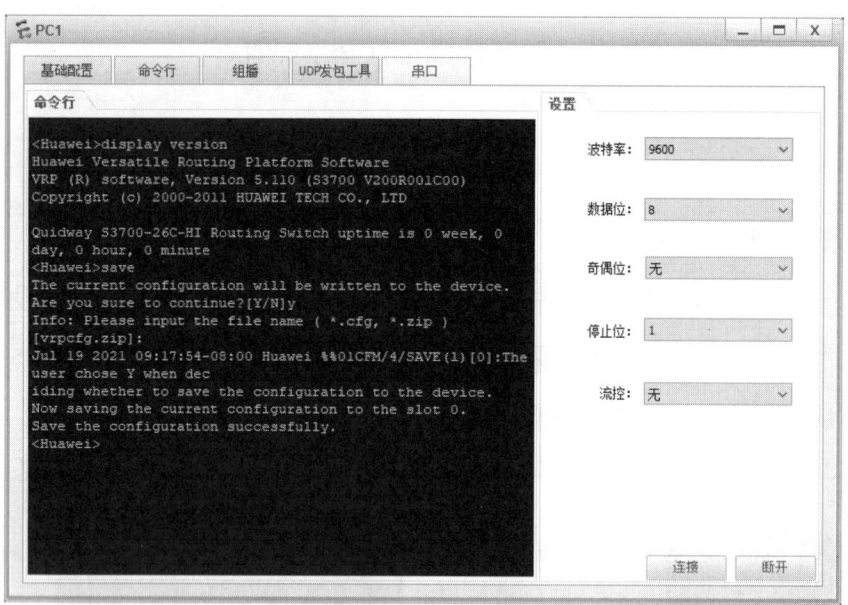

图 5.4 查看交换机的软/硬件版本信息并保存

2. 交换机的基础配置

步骤 1：恢复交换机的出厂设置并重启。

```
<Huawei>reset saved-configuration                        //恢复出厂设置
The configuration will be erased to reconfigure. Continue? [Y/N]:y
<Huawei>reboot                                           //重启交换机
Warning: All the configuration will be saved to the configuration file for the next startup:, Continue?[Y/N]:n
System will reboot! Continue?[Y/N]:y
```

步骤 2：修改交换机的名称。

```
[Huawei]sysname S1                                       //修改交换机的名称
[S1]                                                     //已生效
```

步骤 3：设置交换机所在的时区和系统时间。

```
<S1>clock timezone BJ add 08:00:00                       //设置时区
<S1>clock datetime 11:20:00 2022-11-02                   //设置系统时间
<S1>display clock                                        //查看系统时间
2022-11-02 11:20:25+08:00
Wednesday
Time Zone(BJ) : UTC+08:00
```

步骤 4：设置语言模式。

```
<S1>language-mode Chinese                                //设置语言模式为"中文"
Change language mode, confirm? [Y/N] y
提示：改变语言模式成功。
```

步骤 5：设置交换机远程管理的 IP 地址。

```
[S1]int Vlanif 1                                         //进入 VLAN1
```

```
[S1-Vlanif1]ip add 192.168.1.254 255.255.255.0          //设置VLAN1的IP地址
```

步骤6：取消干扰信息，并设置永不超时。

```
<S1>undo terminal monitor                               //取消干扰信息
Info: Current terminal monitor is off.
<S1>system-view
[S1]user-interface console 0                            //进入Console口
[S1-ui-console0]idle-timeout 0                          //设置永不超时
[S1-ui-console0]quit
```

步骤7：配置交换机的Console口密码。

（1）以登录用户界面的认证方式为"密码认证"，密码为"123456"为例，配置如下。

```
[S1]user-interface console 0                            //进入Console口
[S1-ui-console0]authentication-mode password            //使用密码认证方式
[S1-ui-console0]set authentication password simple 123456  //密码为"123456"
[S1-ui-console0]return
<S1>quit
测试：
Password:                                               //输入密码，此处不显示
<S1>
```

（2）以登录用户界面的认证方式为"AAA认证"，用户名为"admin"，密码为"123456"为例，配置如下。

```
<S1>system-view
[S1]user-interface console 0
[S1-ui-console0]authentication-mode aaa
[S1-ui-console0]quit
[S1]aaa
[S1-aaa]local-user admin password simple 123456         //配置本地用户的登录密码
[S1-aaa]local-user admin service-type terminal          //配置本地用户的接入方式
[S1-aaa]return
<S1>quit
测试：
Username:admin                                          //输入用户名"admin"
Password:                                               //输入密码，此处不显示
<S1>
```

步骤8：撤销交换机配置时的弹出信息。

```
[S1]undo info-center enable                             //撤销配置时的弹出信息
Info: Information center is disabled.
```

步骤9：设置端口带宽。

对交换机的端口进行带宽限制，一般可以将端口带宽设置为10Mbit/s、100Mbit/s和Auto自适应3种，设置的方法很简单，使用"speed"命令即可实现。先使用"undo negotiation auto"命令关闭自协商功能,再设置端口带宽。

```
[S1]int Ethernet 0/0/6                                    //进入端口配置模式
[S1-Ethernet0/0/6]speed ?
  10    10M port speed mode
  100   100M port speed mode
[S1-Ethernet0/0/6]undo   negotiation auto                 //关闭自协商功能
[S1-Ethernet0/0/6]speed 10                                //设置端口带宽为10Mbit/s
[S1-Ethernet0/0/6]quit
```

步骤 10：配置端口双工模式。

```
[S1]int Ethernet 0/0/7
[S1-Ethernet0/0/7]duplex ?
  full  Full-Duplex mode
  half  Half-Duplex mode
[S1-Ethernet0/0/7]undo negotiation auto
[S1-Ethernet0/0/7]duplex full
```

步骤 11：管理 MAC 地址表。

交换机是工作在数据链路层的网络设备，当交换机的端口接入网络设备（如计算机）时，交换机会自动生成 MAC 地址表，查看交换机的 MAC 地址表的命令为"display mac-address"；在交换机中添加静态条目的命令为"mac-address static"；清除 MAC 地址表的命令为"reset arp all"。

（1）在刚启动时，查看交换机的 MAC 地址表。

```
[S1]display mac-address                                   //地址表为空的
[S1]
```

（2）在 PC1 上 ping S1（确保是连通的），并查看交换机的 MAC 地址表。

```
[S1]display mac-address
MAC address table of slot 0:
------------------------------------------------------------------------
MAC Address       VLAN/          PEVLAN CEVLAN Port       Type        LSP/LSR-ID
                  VSI/SI                                              MAC-Tunnel
------------------------------------------------------------------------
5489-98cd-33aa 1    -              -      Eth0/0/1    dynamic     0/-
------------------------------------------------------------------------
Total matching items on slot 0 displayed = 1
```

（3）使用"mac-address static"命令在交换机中添加静态条目，并查看 MAC 地址表的信息。

```
[S1]mac-address static 5489-9865-74db Ethernet 0/0/1 vlan 1
[S1]display mac-address
MAC address table of slot 0:
------------------------------------------------------------------------
MAC Address       VLAN/          PEVLAN CEVLAN Port       Type        LSP/LSR-ID
```

```
                VSI/SI                              MAC-Tunnel
------------------------------------------------------------------
5489-9865-74db 1          -         -     Eth0/0/1 static    -
------------------------------------------------------------------
Total matching items on slot 0 displayed = 1
```

（4）使用"undo"命令清除 MAC 地址表。

```
[S1]undo mac-address static 5489-9865-74db Ethernet 0/0/1 vlan 1
[S1]quit
<S1>display mac-address
<S1>
```

步骤 12：保存设备的当前配置。

```
<S1>save
The current configuration will be written to the device.
Are you sure to continue?[Y/N]y
Info: Please input the file name ( *.cfg, *.zip ) [vrpcfg.zip]:
May 28 2020 11:40:36-08:00 Huawei %%01CFM/4/SAVE(l)[50]:The user chose Y when
deciding whether to save the configuration to the device.
Now saving the current configuration to the slot 0.
Save the configuration successfully.
```

任务验收

（1）交换机都可以通过 Console 口与计算机的 COM 口连接，并进行相应的配置。

（2）测试 Console 口的密码配置是否正确。

（3）使用"display"命令查看交换机的 MAC 地址表中是否存在静态条目。

（4）使用"display current-configuration"命令查看设备的当前配置。

```
[S1]display current-configuration                    //查看设备的当前配置
#
sysname S1
……                                                   //此处省略部分内容
#
local-user admin password simple 123456
 local-user admin service-type terminal
#
interface Vlanif1
 ip address 192.168.1.254 255.255.255.0
#
interface MEth0/0/1
#
interface Ethernet0/0/1
#
```

```
interface Ethernet0/0/2
#
interface Ethernet0/0/3
 undo negotiation auto
#
interface Ethernet0/0/4
 undo negotiation auto
#
......                                          //此处省略部分内容
#
user-interface con 0
 authentication-mode aaa
 idle-timeout 0 0
user-interface vty 0 4
#
return
```

（5）连通性测试。

① 右击 PC1，在弹出的快捷菜单中选择"设置"命令，打开 PC1 的设置对话框。首先在"基础配置"选项卡的"IPv4 配置"选区中，选中"静态"单选按钮，然后设置 IP 地址为 192.168.1.1，子网掩码为 255.255.255.0，如图 5.5 所示，最后单击"应用"按钮。

② 选择设置对话框的"命令行"选项卡，在"PC>"后输入"ping 192.168.1.254"，并按"Enter"键进行测试，连通性测试结果如图 5.6 所示。

图 5.5　PC1 的设置对话框

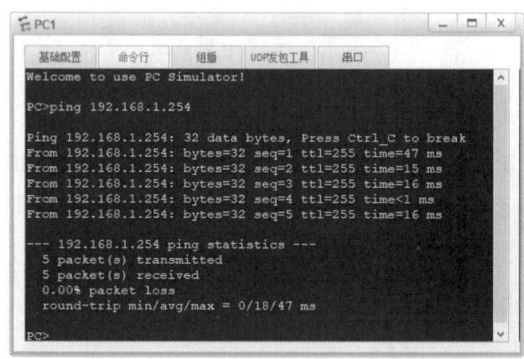

图 5.6　连通性测试结果

任务 6　实现不同部门的计算机之间的网络隔离

任务目标

1. 理解 VLAN 的作用和特点。
2. 学会交换机 VLAN 的划分方法。

任务描述

某公司的局域网搭建完成，网络管理员小赵按照公司的要求，根据部门的不同隔离出多个办公区网络。财务部和销售部都在同一楼层办公，计算机都连接在同一台交换机上。在工作中，由于计算机病毒等可能造成部门之间的计算机交叉感染，部门网络的安全得不到保障。因此，公司要求小赵按照部门划分子网。在二层交换机上无法实现子网的划分，但利用交换机上的 VLAN（Virtual Local Area Network，虚拟局域网）技术和二层技术可以实现三层子网的划分，从而实现不同部门的计算机之间的网络隔离。

任务要求

（1）为了实现不同部门的计算机之间的网络隔离，网络拓扑结构如图 6.1 所示。

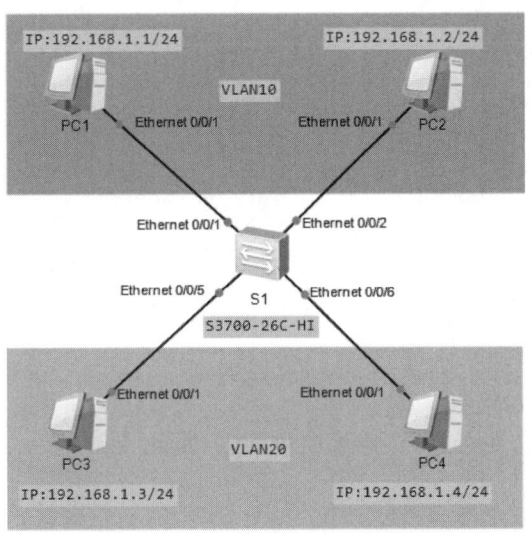

图 6.1　不同部门的计算机之间网络隔离的网络拓扑结构

（2）交换机的 VLAN 划分情况如表 6.1 所示。

表 6.1　交换机的 VLAN 划分情况

VLAN 编号	VLAN 名称	端口范围	连接的计算机	备注
10	Finance	Ethernet 0/0/1～Ethernet 0/0/4	PC1、PC2	财务部
20	Sales	Ethernet 0/0/5～Ethernet 0/0/8	PC3、PC4	销售部

（3）PC1～PC4 的端口及 IP 地址设置如表 6.2 所示。

表 6.2　PC1～PC4 的端口及 IP 地址设置

设备名称	本端端口	IP 地址/子网掩码	网关	对端设备与端口
PC1	Ethernet 0/0/1	192.168.1.1/24	—	S1:Ethernet 0/0/1
PC2	Ethernet 0/0/1	192.168.1.2/24	—	S1:Ethernet 0/0/2
PC3	Ethernet 0/0/1	192.168.1.3/24	—	S1:Ethernet 0/0/5
PC4	Ethernet 0/0/1	192.168.1.4/24	—	S1:Ethernet 0/0/6

（4）通过交换机的 VLAN 划分，实现不同部门的计算机之间的网络隔离。验证是否可以实现接入相同 VLAN 的计算机能相互通信，接入不同 VLAN 的计算机不能相互通信。

任务实施

步骤 1：按照图 6.1 搭建网络拓扑，连线全部使用直通线，开启所有设备电源，并为每台计算机设置相应的 IP 地址和子网掩码。

步骤 2：创建 VLAN。

除默认的 VLAN1，其他 VLAN 需要通过命令来手动创建。创建 VLAN 有两种方式，一种方式是使用"vlan"命令一次创建单个 VLAN，另一种方式是使用"vlan batch"命令一次创建多个 VLAN。

```
<Huawei>system-view                      //进入系统视图
[Huawei]sysname S1                       //修改主机名称
[S1]vlan 10                              //创建 VLAN10
[S1-vlan10]description Finance           //命名 VLAN 为"Finance"
[S1-vlan10]vlan 20
[S1-vlan20]description Sales             //命名 VLAN 为"Sales"
[S1-vlan20]quit                          //退出系统视图
```

知识链接：

交换机 VLAN 的创建需要在系统视图中进行，因此要先进入系统视图。VLAN 的命令很简单。

（1）创建 VLAN：vlan [vlan id]（如 vlan10）。

（2）删除 VLAN：undo vlan [vlan id]（如 undo vlan10）。

（3）如果同时创建 3 个 VLAN，分别为 VLAN10、VLAN20 和 VLAN30，则可以使用"vlan batch"命令创建 VLAN10、VLAN20 和 VLAN30：[Huawei]vlan batch 10 20 30。

步骤 3：查看 VLAN 的相关信息[①]。

```
[S1]display vlan                                          //查看VLAN
The total number of vlans is : 3
--------------------------------------------------------------------------
U: Up;         D: Down;         TG: Tagged;         UT: Untagged;
MP: Vlan-mapping;               ST: Vlan-stacking;
#: ProtocolTransparent-vlan;    *: Management-vlan;
--------------------------------------------------------------------------
VID Type    Ports
--------------------------------------------------------------------------
1   common  UT:Eth0/0/1(D)      Eth0/0/2(D)       Eth0/0/3(D)       Eth0/0/4(D)
            Eth0/0/5(D)         Eth0/0/6(D)       Eth0/0/7(D)       Eth0/0/8(D)
            Eth0/0/9(D)         Eth0/0/10(D)      Eth0/0/11(D)      Eth0/0/12(D)
            Eth0/0/13(D)        Eth0/0/14(D)      Eth0/0/15(D)      Eth0/0/16(D)
            Eth0/0/17(D)        Eth0/0/18(D)      Eth0/0/19(D)      Eth0/0/20(D)
            Eth0/0/21(D)        Eth0/0/22(D)      GE0/0/1(D)        GE0/0/2(D)
10  common
20  common
VID Status  Property      MAC-LRN Statistics Description
--------------------------------------------------------------------------
1   enable  default       enable  disable    VLAN 0001
10  enable  default       enable  disable    Finance
20  enable  default       enable  disable    Sales
//可以观察到，S1已经成功创建了相应VLAN，但目前没有任何端口加入创建的VLAN10和VLAN20中，在默认情
况下交换机上的所有端口都属于VLAN1
```

步骤 4：配置 Access 及划分 VLAN 端口。

```
[S1]interface Ethernet 0/0/1                   //进入Ethernet 0/0/1端口
[S1-Ethernet0/0/1]port link-type access        //将端口配置为Access
[S1-Ethernet0/0/1]port default vlan 10         //将端口划分到VLAN10中
[S1-Ethernet0/0/1]quit
[S1]interface Ethernet 0/0/2                   //进入Ethernet 0/0/2端口
[S1-Ethernet0/0/2]port link-type access        //将端口配置为Access
[S1-Ethernet0/0/2]port default vlan 10         //将端口划分到VLAN10中
[S1-Ethernet0/0/2]quit
[S1]interface Ethernet 0/0/3                   //进入Ethernet 0/0/3端口
[S1-Ethernet0/0/3]port link-type access        //将端口配置为Access
```

---

① 在本书中，Eth 表示 Ethernet。

```
[S1-Ethernet0/0/3]port default vlan 10                    //将端口划分到VLAN10中
[S1-Ethernet0/0/3]quit
[S1]interface Ethernet 0/0/4
[S1-Ethernet0/0/4]port link-type access
[S1-Ethernet0/0/4]port default vlan 10
[S1-Ethernet0/0/4]quit
[S1]port-group 1
[S1-port-group-1]group-member Ethernet 0/0/5 to Ethernet 0/0/8  //将多个端口添加到端口组
[S1-port-group-1]port link-type access                    //将多个端口配置为Access
[S1-Ethernet0/0/5]port link-type access
[S1-Ethernet0/0/6]port link-type access
[S1-Ethernet0/0/7]port link-type access
[S1-Ethernet0/0/8]port link-type access
[S1-port-group-1]port default vlan 20                     //将多个端口划分到VLAN20中
[S1-Ethernet0/0/5]port default vlan 20
[S1-Ethernet0/0/6]port default vlan 20
[S1-Ethernet0/0/7]port default vlan 20
[S1-Ethernet0/0/8]port default vlan 20
[S1-port-group-1]quit
```

步骤5：再次查看VLAN的相关信息。

```
[S1]display vlan
The total number of vlans is : 3
--------------------------------------------------------------------------------
U: Up;         D: Down;        TG: Tagged;        UT: Untagged;
MP: Vlan-mapping;              ST: Vlan-stacking;
#: ProtocolTransparent-vlan;   *: Management-vlan;
--------------------------------------------------------------------------------
VID  Type    Ports
--------------------------------------------------------------------------------
1    common  UT:Eth0/0/9(D)    Eth0/0/10(D)    Eth0/0/11(D)    Eth0/0/12(D)
             Eth0/0/13(D)      Eth0/0/14(D)    Eth0/0/15(D)    Eth0/0/16(D)
             Eth0/0/17(D)      Eth0/0/18(D)    Eth0/0/19(D)    Eth0/0/20(D)
             Eth0/0/21(D)      Eth0/0/22(D)    GE0/0/1(D)      GE0/0/2(D)
10   common  UT:Eth0/0/1(U)    Eth0/0/2(U)     Eth0/0/3(D)     Eth0/0/4(D)
20   common  UT:Eth0/0/5(U)    Eth0/0/6(U)     Eth0/0/7(D)     Eth0/0/8(D)
VID  Status  Property     MAC-LRN Statistics Description
--------------------------------------------------------------------------------
1    enable  default      enable  disable    VLAN 0001
10   enable  default      enable  disable    Finance
20   enable  default      enable  disable    Sales
//通过以上操作，在交换机上进行了VLAN的创建和端口的划分，从而实现了交换机端口的隔离
```

任务验收

（1）确认计算机已经正确连接到对应 VLAN 端口。例如，在 PC1、PC2 连接 VLAN10 时，只能接入交换机的 Ethernet 0/0/1～Ethernet 0/0/4 端口。

（2）先使用相同 VLAN 的计算机进行"ping"命令测试，再使用不同 VLAN 之间的计算机进行"ping"命令测试。下面分别使用 PC1 与 PC2、PC3 进行"ping"命令测试，连通性测试结果如图 6.2 所示。

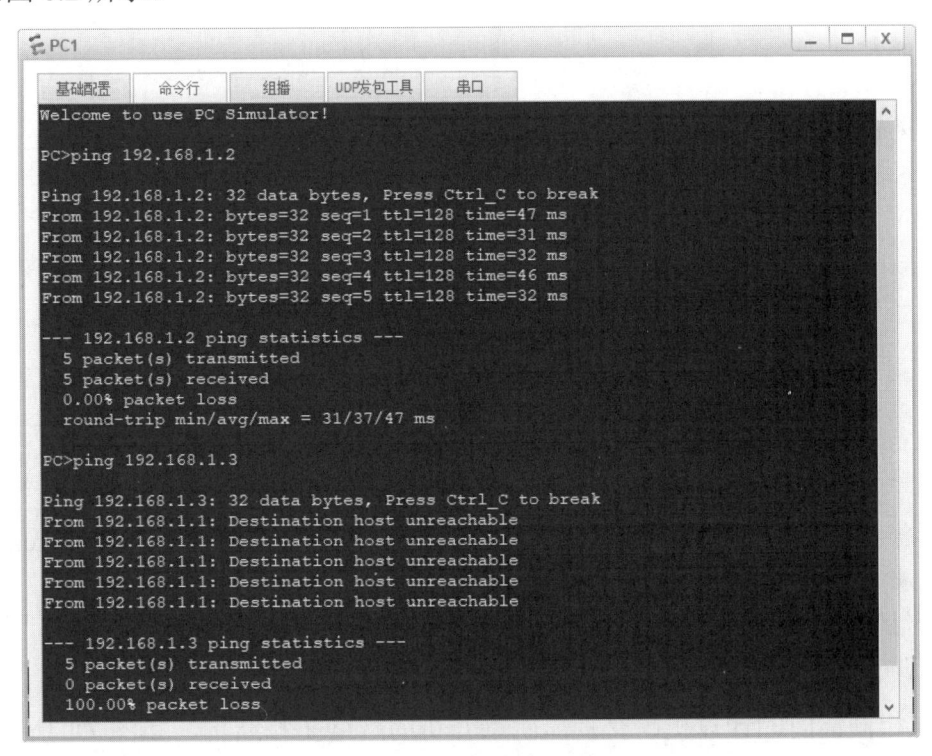

图 6.2　连通性测试结果

任务 7　实现相同部门的计算机跨交换机的网络互访

任务目标

1. 理解交换机中继链路的作用。
2. 学会配置交换机间相同 VLAN 通信方法。

任务描述

某公司有财务部、市场部等部门，其中在不同楼层内都有财务部和市场部的计算机。为了更加安全与便捷地管理公司，网络管理员小赵通过划分 VLAN 使财务部和市场部的计算机之间不可以自由通信，分布在不同楼层的交换机上相同部门的计算机可以相互通信，这就需要使用 802.1q 协议实现相同部门的计算机跨交换机的网络互访，也就是在两个交换机之间开启 Trunk 通信。

任务要求

（1）实现相同部门的计算机跨交换机网络互访的网络拓扑结构，如图 7.1 所示。

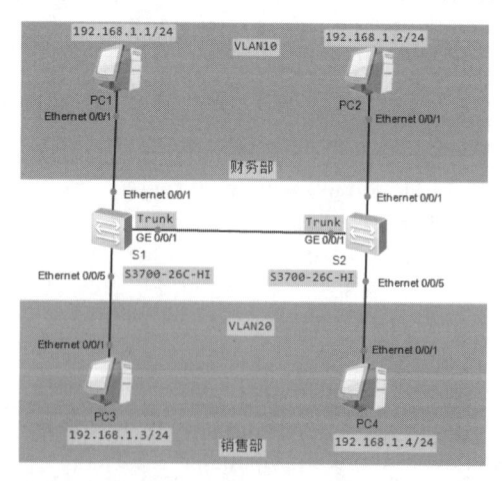

图 7.1　实现相同部门的计算机跨交换机网络互访的网络拓扑结构

（2）在交换机 S1 和交换机 S2 上分别划分两个 VLAN（VLAN10 和 VLAN20），划分情况如表 7.1 所示。

表 7.1 交换机的 VLAN 划分情况

VLAN 编号	VLAN 名称	端口范围	连接的计算机	备注
10	Finance	Ethernet 0/0/1～Ethernet 0/0/4	PC1、PC2	财务部
20	Sales	Ethernet 0/0/5～Ethernet 0/0/8	PC3、PC4	销售部
Trunk	—	GE 0/0/1	—	—

（3）PC1～PC4 的端口及 IP 地址设置如表 7.2 所示。

表 7.2　PC1～PC4 的端口及 IP 地址设置

设备名称	本端端口	IP 地址/子网掩码	网关	对端设备与端口
PC1	Ethernet 0/0/1	192.168.1.1/24	—	S1:Ethernet 0/0/1
PC2	Ethernet 0/0/1	192.168.1.2/24	—	S2:Ethernet 0/0/1
PC3	Ethernet 0/0/1	192.168.1.3/24	—	S1:Ethernet 0/0/5
PC4	Ethernet 0/0/1	192.168.1.4/24	—	S2:Ethernet 0/0/5

（4）通过配置交换机相同 VLAN 内的计算机相互通信，实现相同部门计算机的网络互访，即实现 PC1 与 PC3 相互通信，PC2 与 PC4 相互通信，其他组合不能通信。

任务实施

步骤 1：按照图 7.1 搭建网络拓扑，连线全部使用直通线，开启所有设备电源，并为每台计算机设置相应的 IP 地址和子网掩码。

步骤 2：设置交换机的名称，创建 VLAN，并配置 Access 及划分端口。

对两个交换机进行相同的 VLAN 划分，下面是交换机 S1 的配置过程，按照相同方式可实现交换机 S2 的配置[①]。

```
<Huawei>system-view
[Huawei]sysname S1
[S1]vlan 10
[S1-vlan10]description Finance
[S1-vlan10]vlan 20
[S1-vlan20]description Sales
[S1-vlan20]quit
[S1]port-group 1
```

---

① 在本书中，e 表示 Ethernet。

```
[S1-port-group-1]group-member e0/0/1 to e0/0/4
[S1-port-group-1]port link-type access
[S1-Ethernet0/0/1]port link-type access
[S1-Ethernet0/0/2]port link-type access
[S1-Ethernet0/0/3]port link-type access
[S1-Ethernet0/0/4]port link-type access
[S1-port-group-1]port default vlan 10
[S1-Ethernet0/0/1]port default vlan 10
[S1-Ethernet0/0/2]port default vlan 10
[S1-Ethernet0/0/3]port default vlan 10
[S1-Ethernet0/0/4]port default vlan 10
[S1-port-group-1]quit
[S1]port-group 2
[S1-port-group-2]group-member e0/0/5 to e0/0/8
[S1-port-group-2]port link-type access
[S1-Ethernet0/0/5]port link-type access
[S1-Ethernet0/0/6]port link-type access
[S1-Ethernet0/0/7]port link-type access
[S1-Ethernet0/0/8]port link-type access
[S1-port-group-2]port default vlan 20
[S1-Ethernet0/0/5]port default vlan 20
[S1-Ethernet0/0/6]port default vlan 20
[S1-Ethernet0/0/7]port default vlan 20
[S1-Ethernet0/0/8]port default vlan 20
[S1-port-group-2]quit
```

步骤3：查看交换机 S1 的 VLAN 配置。

```
[S1]display vlan
The total number of vlans is : 3
--------------------------------------------------------------------
VID  Type    Ports
--------------------------------------------------------------------
1    common  UT:Eth0/0/9(D)    Eth0/0/10(D)   Eth0/0/11(D)   Eth0/0/12(D)
                Eth0/0/13(D)   Eth0/0/14(D)   Eth0/0/15(D)   Eth0/0/16(D)
                Eth0/0/17(D)   Eth0/0/18(D)   Eth0/0/19(D)   Eth0/0/20(D)
                Eth0/0/21(D)   Eth0/0/22(D)   GE0/0/1(U)     GE0/0/2(D)
10   common  UT:Eth0/0/1(U)    Eth0/0/2(D)    Eth0/0/3(D)    Eth0/0/4(D)
20   common  UT:Eth0/0/5(U)    Eth0/0/6(D)    Eth0/0/7(D)    Eth0/0/8(D)
VID  Status  Property    MAC-LRN Statistics Description
--------------------------------------------------------------------
1    enable  default     enable  disable    VLAN 0001
10   enable  default     enable  disable    Finance
20   enable  default     enable  disable    Sales
```

步骤 4：查看交换机 S2 的 VLAN 配置。

```
[S2]display vlan
The total number of vlans is : 3
--------------------------------------------------------------------------------
VID  Type    Ports
--------------------------------------------------------------------------------
1    common  UT:Eth0/0/9(D)    Eth0/0/10(D)   Eth0/0/11(D)   Eth0/0/12(D)
             Eth0/0/13(D)      Eth0/0/14(D)   Eth0/0/15(D)   Eth0/0/16(D)
             Eth0/0/17(D)      Eth0/0/18(D)   Eth0/0/19(D)   Eth0/0/20(D)
             Eth0/0/21(D)      Eth0/0/22(D)   GE0/0/1(U)     GE0/0/2(D)
10   common  UT:Eth0/0/1(U)    Eth0/0/2(D)    Eth0/0/3(D)    Eth0/0/4(D)
20   common  UT:Eth0/0/5(U)    Eth0/0/6(D)    Eth0/0/7(D)    Eth0/0/8(D)
VID  Status  Property     MAC-LRN Statistics Description
--------------------------------------------------------------------------------
1    enable  default      enable  disable    VLAN 0001
10   enable  default      enable  disable    Finance
20   enable  default      enable  disable    Sales
```

当两台交换机都按照上面的命令完成配置后，进行测试可以发现，现在 4 台计算机都不能相互通信了。寻找原因，发现交换机是通过 GE 0/0/1 端口进行连接的，而 GE 0/0/1 端口并不在 VLAN10 和 VLAN20 中。可以尝试先将与交换机连接的 GE 0/0/1 端口改为与 Ethernet 0/0/2 端口（VLAN10 的端口）连接，再次测试可以发现 PC1 和 PC2 能够相互通信了，而 PC3 和 PC4 仍然不能相互通信。同样，将交换机与 VLAN20 的端口连接，PC3 和 PC4 可以相互通信。

步骤 5：配置 Trunk 端口和允许通过的 VLAN。

要解决上述难题，仍然通过 GE 0/0/1 端口连接两台交换机，可以先将 GE 0/0/1 端口配置为 Trunk，再在该 Trunk 链路上配置允许通过的单个、多个 VLAN，或者在交换机上配置允许所有的 VLAN 通过。

```
[S1]interface GigabitEthernet 0/0/1
[S1-GigabitEthernet0/0/1]port link-type trunk              //将端口配置为Trunk
[S1-GigabitEthernet0/0/1]port trunk allow-pass vlan ?
 INTEGER<1-4094>  VLAN ID                                  //允许通过的VLAN的ID
 all              All                                      //允许所有的VLAN通过
[S1-GigabitEthernet0/0/1]port trunk allow-pass vlan 10 20
//允许VLAN10和VLAN20通过
```

> **小贴士**
>
> 在配置华为交换机时，要明确配置被允许通过的 VLAN，从而实现对 VLAN 流量转发的控制。

步骤 6：使用"display vlan"命令查看端口模式，GE 0/0/1 端口的链路类型为 TG，说明已经是 Trunk 链路状态了。

```
[S1]display vlan
The total number of vlans is : 3
--------------------------------------------------------------------
U: Up;         D: Down;         TG: Tagged;       UT: Untagged;
MP: Vlan-mapping;               ST: Vlan-stacking;
#: ProtocolTransparent-vlan;    *: Management-vlan;
--------------------------------------------------------------------
VID  Type    Ports
--------------------------------------------------------------------
10   common  UT:Eth0/0/1(U)   Eth0/0/2(D)   Eth0/0/3(D)   Eth0/0/4(D)
             TG:GE0/0/1(U)
20   common  UT:Eth0/0/5(U)   Eth0/0/6(D)   Eth0/0/7(D)   Eth0/0/8(D)
             TG:GE0/0/1(U)
```

同理，可以配置交换机 S2 的 GE 0/0/1 端口为 Trunk，并配置允许 VLAN10 和 VLAN20 通过。至此，本任务配置完成。这时两台相同 VLAN 中的计算机已经可以相互通信了。

步骤 7：检查 GE 0/0/1 端口 Trunk 的配置情况。

使用"display port vlan GigabitEthernet 0/0/1"命令查看端口模式，GE 0/0/1 端口的链路类型为 Trunk，并允许 VLAN10 和 VLAN20 通过。

（1）在交换机 S1 上查看端口模式。

```
[S1]display port vlan GigabitEthernet 0/0/1
Port                    Link Type    PVID   Trunk VLAN List
--------------------------------------------------------------------
GigabitEthernet0/0/1    trunk        1      1 10 20
```

（2）在交换机 S2 上查看端口模式。

```
[S1]display port vlan GigabitEthernet 0/0/1
Port                    Link Type    PVID   Trunk VLAN List
--------------------------------------------------------------------
GigabitEthernet0/0/1    trunk        1      1 10 20
```

任务验收

在 PC1 上 ping PC2 的 IP 地址"192.168.1.2"，如果 PC1 与 PC2 是连通的，则表明交换机之前的 Trunk 链路已经成功建立。在 PC1 上 ping PC4 的 IP 地址"192.168.1.4"，如果 PC1 与 PC4 不是连通的，则表明不同 VLAN 之间的计算机无法相互通信，连通性测试结果如图 7.2 所示。

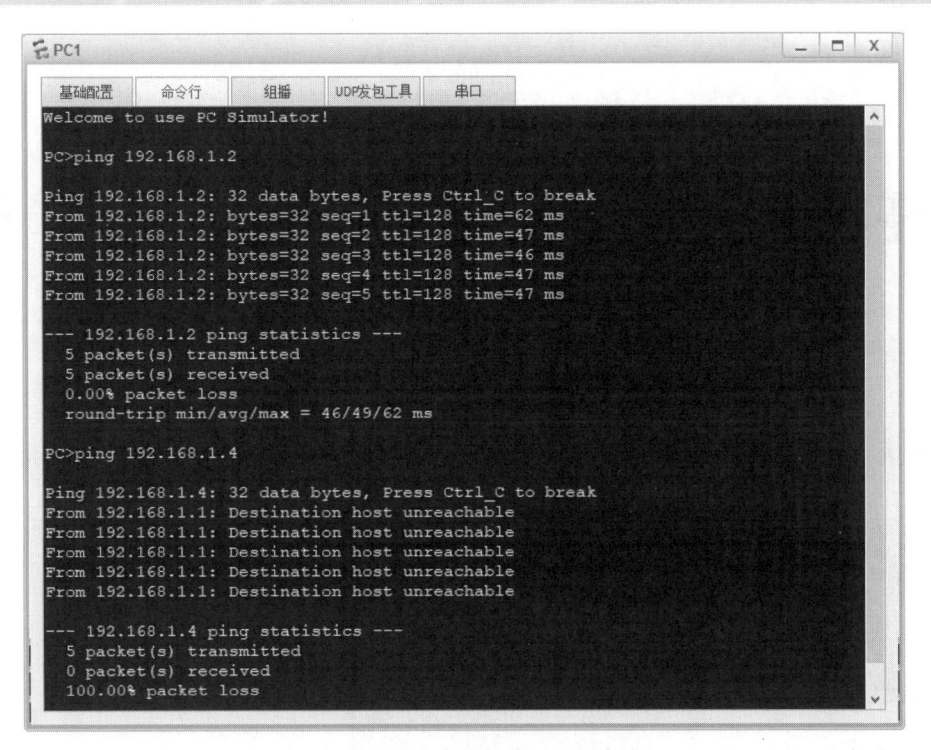

图 7.2 连通性测试结果

任务 8　利用三层交换机实现部门计算机之间的网络互访

任务目标

1．理解三层交换机的功能和作用。
2．学会利用三层交换机实现不同 VLAN 之间计算机通信的方法。

任务描述

某公司按照部门业务不同，划分出多个不同 VLAN，从而实现部门计算机之间的网络隔离。隔离后的部门网络虽然暂时解决了安全和干扰问题，但也会导致不同 VLAN 的计算机之间不能相互通信，公司内部的公共资源不能共享。因此，网络管理员小赵需要实现所有部门计算机之间的网络互访。

任务要求

在交换机中利用 VLAN 中继技术，可以实现同一部门 VLAN 内计算机的跨交换机通信。如果要实现不同 VLAN 之间计算机的网络互访，则需要利用三层交换机技术。

（1）利用三层交换机实现部门计算机之间的网络互访，网络拓扑结构如图 8.1 所示。

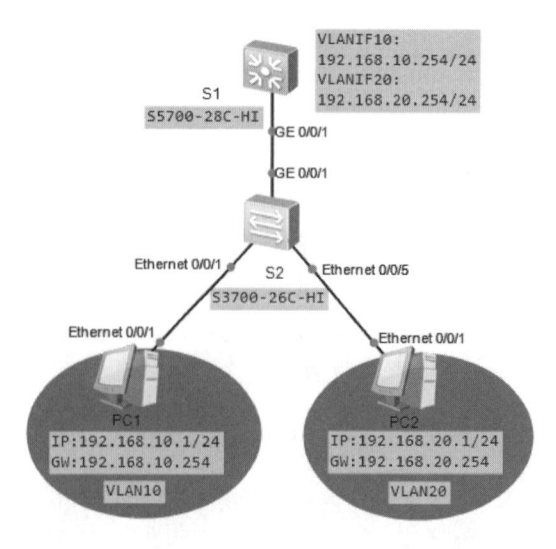

图 8.1　利用三层交换机实现部门计算机之间网络互访的网络拓扑结构

（2）在交换机 S1 和交换机 S2 上分别划分两个 VLAN（VLAN10、VLAN20），并将 GE 0/0/1 端口设置为 Trunk，划分情况如表 8.1 所示。

表 8.1 交换机的 VLAN 划分情况

设备名称	VLAN 编号	端口范围	IP 地址/端口模式
S1	10	—	192.168.10.254/24
	20	—	192.168.20.254/24
	—	GE 0/0/1	Trunk
S2	10	Ethernet 0/0/1～Ethernet 0/0/4	Access
	20	Ethernet 0/0/5～Ethernet 0/0/8	Access
	—	GE 0/0/1	Trunk

（3）PC1、PC2 的端口及 IP 地址设置如表 8.2 所示。

表 8.2 PC1、PC2 的端口及 IP 地址设置

设备名称	本端端口	IP 地址/子网掩码	网关	所属 VLAN	对端设备与端口
PC1	Ethernet 0/0/1	192.168.10.1/24	192.168.10.254	10	S2:Ethernet 0/0/1
PC2	Ethernet 0/0/1	192.168.20.1/24	192.168.20.254	20	S2:Ethernet 0/0/5

（4）通过三层交换机实现不同VLAN的计算机之间的通信。

任务实施

步骤 1：按照图 8.1 搭建网络拓扑，连线全部使用直通线，开启所有设备电源，并为每台计算机设置相应的 IP 地址和子网掩码。

步骤 2：划分二层交换机的 VLAN 及端口。

```
<Huawei>system-view
[Huawei]sysname S1
[S1]vlan batch 10 20                                          //创建VLAN10和VLAN20
[S1]port-group 1
[S1-port-group-1]group-member Ethernet 0/0/1 to Ethernet 0/0/4
[S1-port-group-1]port link-type access
[S1-Ethernet0/0/1]port link-type access
[S1-Ethernet0/0/2]port link-type access
[S1-Ethernet0/0/3]port link-type access
[S1-Ethernet0/0/4]port link-type access
[S1-port-group-1]port default vlan 10
[S1-Ethernet0/0/1]port default vlan 10
[S1-Ethernet0/0/2]port default vlan 10
[S1-Ethernet0/0/3]port default vlan 10
[S1-Ethernet0/0/4]port default vlan 10
[S1-port-group-1]quit
```

```
[S1]port-group 2
[S1-port-group-2]group-member Ethernet 0/0/5 to Ethernet 0/0/8
[S1-port-group-2]port link-type access
[S1-Ethernet0/0/5]port link-type access
[S1-Ethernet0/0/6]port link-type access
[S1-Ethernet0/0/7]port link-type access
[S1-Ethernet0/0/8]port link-type access
[S1-port-group-2]port default vlan 20
[S1-Ethernet0/0/5]port default vlan 20
[S1-Ethernet0/0/6]port default vlan 20
[S1-Ethernet0/0/7]port default vlan 20
[S1-Ethernet0/0/8]port default vlan 20
[S1-port-group-2]quit
[S1]interface GigabitEthernet 0/0/1
[S1-GigabitEthernet0/0/1]port link-type trunk
[S1-GigabitEthernet0/0/1]port trunk allow-pass vlan 10 20
```

步骤 3：划分三层交换机上的 VLAN、设置每个 VLAN 的端口 IP 地址。

```
<Huawei>system-view
[Huawei]sysname S2
[S2]vlan batch 10 20
[S2]interface Vlanif 10                                    //进入 VLAN10
[S2-Vlanif10]ip add 192.168.10.254 24                      //设置 IP 地址
[S2-Vlanif10]quit
[S2]interface Vlanif 20                                    //进入 VLAN20
[S2-Vlanif20]ip add 192.168.20.254 24                      //设置 IP 地址
[S2-Vlanif20]quit
[S2]interface GigabitEthernet 0/0/1
[S2-GigabitEthernet0/0/1]port link-type trunk
[S2-GigabitEthernet0/0/1]port trunk allow-pass vlan 10 20
```

步骤 4：配置交换机的 Trunk 链路。

（1）在交换机 S1 中进行配置。

```
[S1]interface GigabitEthernet 0/0/1                                //进入 GE 0/0/1 端口
[S1-GigabitEthernet0/0/1]port link-type trunk                      //将端口配置为 Trunk
[S1-GigabitEthernet0/0/1]port trunk allow-pass vlan 10 20   //允许 VLAN10 和 VLAN20 通过
```

（2）在交换机 S2 中进行配置。

```
[S2]interface GigabitEthernet 0/0/1
[S2-GigabitEthernet0/0/1]port link-type trunk
[S2-GigabitEthernet0/0/1]port trunk allow-pass vlan 10 20
```

步骤 5：配置计算机的网关，实现不同 VLAN 之间和不同网络之间的通信。

当计算机跨网络连接时，必须通过网关进行路由转发，因此不仅要配置交换机 VLAN 之间的路由，还要为每台计算机配置网关。

在配置计算机的网关时,应该将其配置为该计算机上连设备的 IP 地址,即下一跳地址。在本任务的拓扑结构中,PC1 的上连设备为交换机 S2 的 VLAN10,VLAN10 的端口 IP 地址为 192.168.10.254,则 VLAN10 的端口 IP 地址为 PC1 的下一跳地址。因此,应该将 PC1 的网关配置为 192.168.10.254。同理,应该将 PC2 的网关配置为 VLAN20 的端口 IP 地址 192.168.20.254。

网关的配置在计算机设置对话框的"IPv4 配置"选区中完成。图 8.2 所示为配置的 PC1 的网关。

图 8.2 配置的 PC1 的网关

> **小贴士**
>
> 切记在配置完 IP 地址等信息后,要单击"应用"按钮,否则不会生效。

使用同样的方法配置 PC2 的网关。至此,本任务的所有配置都已经完成。下面进行验证及测试。

任务验收

在 PC1 上 ping PC2 的 IP 地址"192.168.20.1",可以发现网络是连通的,连通性测试结果如图 8.3 所示。

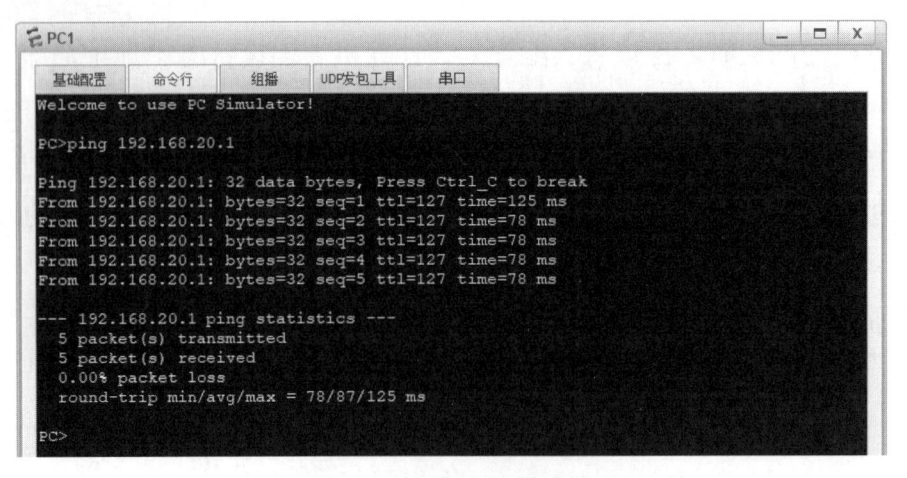

图 8.3　连通性测试结果

任务 9　提高骨干链路的带宽

任务目标

1．理解链路聚合的作用。
2．掌握配置交换机链路聚合技术。

任务描述

某公司的网络中心为了保证接入网络的稳定性，在汇聚层交换机的连接链路上使用多条冗余链路。同时，为了增加带宽，在多条冗余链路之间实现端口聚合，提高骨干链路的带宽，这样可以实现链路之间的冗余和备份效果，从而避免因骨干链路上的单点故障导致的网络中断。

任务要求

（1）提高骨干链路带宽的网络拓扑结构如图 9.1 所示。

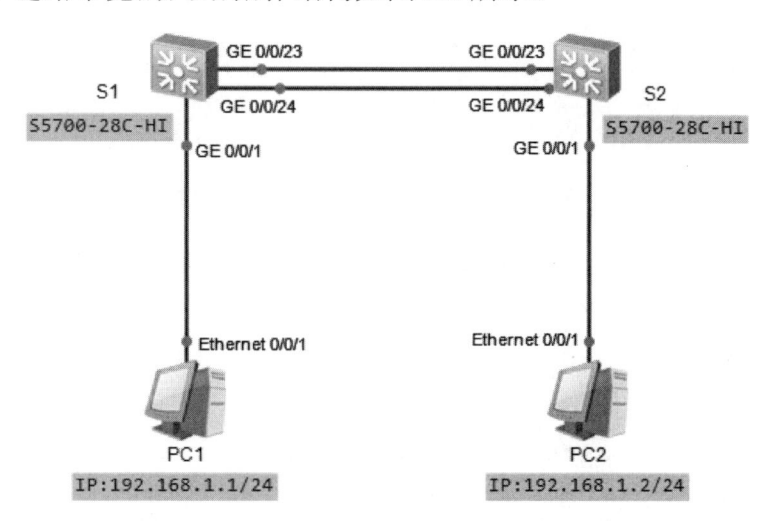

图 9.1　提高主干链路带宽的网络拓扑结构

（2）PC1、PC2 的端口及 IP 地址设置如表 9.1 所示。

表 9.1 PC1、PC2 的端口及 IP 地址设置

设 备 名 称	本 端 端 口	IP 地址/子网掩码	对端设备与端口
PC1	Ethernet 0/0/1	192.168.1.1/24	S1:GE 0/0/1
PC2	Ethernet 0/0/1	192.168.1.2/24	S2:GE 0/0/1

（3）分别设置两台交换机的 GE 0/0/23、GE 0/0/24 端口为端口汇聚，从而实现链路聚合功能，提高骨干链路带宽。

任务实施

步骤 1：按照图 9.1 搭建网络拓扑，连线全部使用直通线，开启所有设备电源，并为每台计算机设置相应的 IP 地址和子网掩码。

步骤 2：配置交换机 S1。

```
<Huawei>system-view
[Huawei]sysname S1
[S1]interface Eth-Trunk 1                    //创建 ID 为 1 的 Eth-Trunk 聚合端口
[S1-Eth-Trunk1]quit                          //退出 Eth-Trunk1 端口视图
[S1]interface GigabitEthernet 0/0/23         //进入 GE 0/0/23 端口视图
[S1-GigabitEthernet0/0/23]eth-trunk 1        //加入 Eth-Trunk 1 聚合端口
[S1-GigabitEthernet0/0/23]quit               //退出 GE 0/0/23 端口视图
[S1]interface GigabitEthernet 0/0/24         //进入 GE 0/0/24 端口视图
[S1-GigabitEthernet0/0/24]eth-trunk 1        //加入 Eth-Trunk 1 聚合端口
[S1-GigabitEthernet0/0/24]quit               //退出 GE 0/0/24 端口视图
[S1]interface Eth-Trunk 1                    //进入 Eth-Trunk 1 聚合端口
[S1-Eth-Trunk1]port link-type trunk          //将端口配置为 Trunk
[S1-Eth-Trunk1]quit                          //退出 Eth-Trunk 1 端口视图
```

步骤3：配置交换机S2。

```
[S2]interface Eth-Trunk 1                    //创建 ID 为 1 的 Eth-Trunk 聚合端口
[S2-Eth-Trunk1]trunkport GigabitEthernet 0/0/23 to 0/0/24
                                             //将 GE 0/0/23 和 GE 0/0/24 端口加入 Eth-Trunk 1
[S2-Eth-Trunk1]port link-type trunk          //将端口配置为 Trunk
[S2-Eth-Trunk1]quit                          //退出 Eth-Trunk 1 端口视图
//这里交换机 S2 使用的是将成员端口批量加入聚合组
```

步骤4：在交换机S1上查看链路聚合组1的信息。

```
[S1]display eth-trunk 1
Eth-Trunk1's state information is:
WorkingMode: NORMAL         Hash arithmetic: According to SIP-XOR-DIP
Least Active-linknumber: 1  Max Bandwidth-affected-linknumber: 8
Operate status: up          Number Of Up Port In Trunk: 2
--------------------------------------------------------------------------
PortName                    Status        Weight
```

GigabitEthernet0/0/23	Up	1
GigabitEthernet0/0/24	Up	1

步骤 5：在交换机 S2 上查看链路聚合组 1 的信息。

```
[S2]display eth-trunk 1
Eth-Trunk1's state information is:
WorkingMode: NORMAL          Hash arithmetic: According to SIP-XOR-DIP
Least Active-linknumber: 1   Max Bandwidth-affected-linknumber: 8
Operate status: up           Number Of Up Port In Trunk: 2
--------------------------------------------------------------------------------
PortName                     Status      Weight
GigabitEthernet0/0/23        Up          1
GigabitEthernet0/0/24        Up          1
//在查看到的信息中表明 Eth-Trunk 工作正常，成员端口都已正确加入
```

任务验收

（1）测试计算机的连通性。在 PC1 上测试 PC1 与 PC2 的连通性，如图 9.2 所示。

（2）修改拓扑结构，并重新测试。去掉一条聚合端口的连线（将其所有的端口关闭即可），重新测试 PC1 与 PC2 的连通性。可以发现去掉一条连线后，计算机的连通性没有受到影响（会出现短暂的丢包情况），连通性测试结果如图 9.3 所示。

图 9.2　测试 PC1 与 PC2 的连通性　　　　　图 9.3　连通性测试结果

任务 10　实现网络负载均衡

任务目标

1．理解交换机生成树的作用和原理。
2．配置多实例生成树协议，从而实现网络负载均衡。

任务描述

某公司最近由于业务迅速发展和对网络可靠性的要求，需要使用两台高性能交换机作为核心交换机，将接入层交换机与核心层交换机连接，形成冗余结构，从而满足网络的可靠性。

利用 STP（Spanning Tree Protocol，生成树协议）和 RSTP（Rapid Spanning Tree Protocol，快速生成树协议）在整个交换网络中生成一个树形拓扑，所有 VLAN 共享一棵生成树，这种结构不能实现网络中流量的负载均衡，反而会导致网络传输过程中的一些交换设备工作繁忙，而另一些交换设备很空闲。配置基于 VLAN 的 MISTP（Multiple Instances Spanning Tree Protocol，多实例生成树协议）不仅可以适应多 VLAN 的场景，还可以在冗余的网络中保障网络的稳健性，并实现网络负载均衡。

任务要求

（1）实现网络负载均衡的网络拓扑结构如图 10.1 所示。
（2）分别在交换机 S1、S2、S3 和 S4 上划分 VLAN10 和 VLAN20，划分情况如表 10.1 所示。

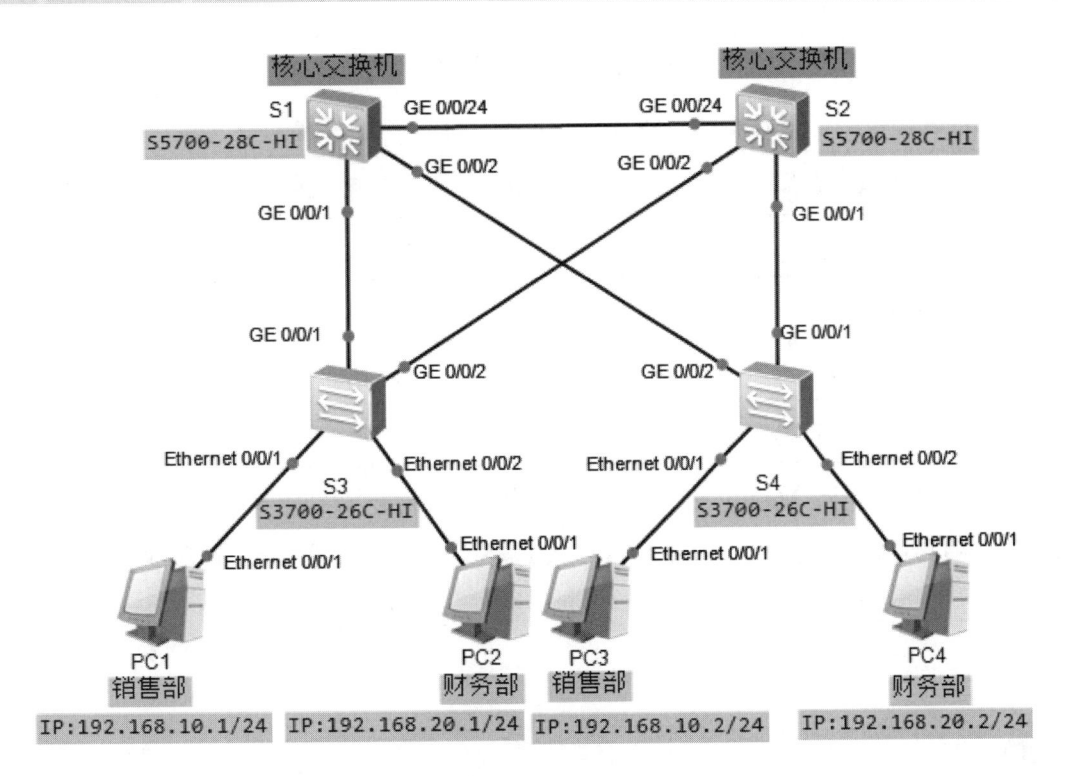

图 10.1　实现网络负载均衡的网络拓扑结构

表 10.1　交换机的 VLAN 划分情况

设 备 名 称	VLAN 编号	端 口 范 围
S1	10	—
	20	—
	Trunk	GE 0/0/1
		GE 0/0/2
		GE 0/0/24
S2	10	—
	20	—
	Trunk	GE 0/0/1
		GE 0/0/2
		GE 0/0/24
S3	10	Ethernet 0/0/1
	20	Ethernet 0/0/2
	Trunk	GE 0/0/1
		GE 0/0/2
S4	10	Ethernet 0/0/1
	20	Ethernet 0/0/2
	Trunk	GE 0/0/1
		GE 0/0/2

（3）PC1～PC4 的端口及 IP 地址设置如表 10.2 所示。

表 10.2　PC1～PC4 的端口 IP 地址设置

设备名称	本地端口	IP 地址/子网掩码	所属 VLAN	对端设备与端口
PC1	Ethernet 0/0/1	192.168.10.1/24	10	S3:Ethernet 0/0/1
PC2	Ethernet 0/0/1	192.168.20.1/24	20	S3:Ethernet 0/0/2
PC3	Ethernet 0/0/1	192.168.10.2/24	10	S4:Ethernet 0/0/1
PC4	Ethernet 0/0/1	192.168.20.2/24	20	S4:Ethernet 0/0/2

（4）在交换机上配置 MSTP，实现网络负载均衡。要求核心交换机有较高优先级，交换机 S1 为主根桥、交换机 S2 为次根桥，VLAN10 的业务流量能够在 S3—S1 链路中传输；交换机 S2 为主根桥、交换机 S1 为次根桥，VLAN20 的业务流量能够在 S4—S2 链路中传输。

任务实施

步骤 1：按照图 10.1 搭建网络拓扑，连线全部使用直通线，开启所有设备电源，并为每台计算机设置相应的 IP 地址和子网掩码。

步骤 2：对交换机进行基本配置。

（1）对交换机 S1 进行基本配置[①]。

```
<Huawei>system-view
[Huawei]sysname S1
[S1]vlan 10
[S1-vlan10]description Sales
[S1-vlan10]quit
[S1]vlan 20
[S1-vlan20]description Finances
[S1-vlan20]quit
[S1]port-group group-member G0/0/1 to G0/0/2 G0/0/24
[S1-port-group]port link-type trunk
[S1-GigabitEthernet0/0/1]port link-type trunk
[S1-GigabitEthernet0/0/2]port link-type trunk
[S1-GigabitEthernet0/0/24]port link-type trunk
[S1-port-group]port trunk allow-pass vlan 10 20
[S1-GigabitEthernet0/0/1]port trunk allow-pass vlan 10 20
[S1-GigabitEthernet0/0/2]port trunk allow-pass vlan 10 20
```

---

① 在本书中，G 表示 GigabitEthernet。

```
[S1-GigabitEthernet0/0/24]port trunk allow-pass vlan 10 20
```

（2）对交换机 S2 进行基本配置。

```
<Huawei>system-view
[Huawei]sysname S2
[S2]vlan 10
[S2-vlan10]description Sales
[S2-vlan10]quit
[S2]vlan 20
[S2-vlan20]description Finances
[S2-vlan20]quit
[S2]port-group group-member G0/0/1 to G0/0/2 G0/0/24
[S2-port-group]port link-type trunk
[S2-GigabitEthernet0/0/1]port link-type trunk
[S2-GigabitEthernet0/0/2]port link-type trunk
[S2-GigabitEthernet0/0/24]port link-type trunk
[S2-port-group]port trunk allow-pass vlan 10 20
[S2-GigabitEthernet0/0/1]port trunk allow-pass vlan 10 20
[S2-GigabitEthernet0/0/2]port trunk allow-pass vlan 10 20
[S2-GigabitEthernet0/0/24]port trunk allow-pass vlan 10 20
```

（3）对交换机 S3 进行基本配置。

```
<Huawei>system-view
[Huawei]sysname S3
[S3]vlan 10
[S3-vlan10]description Sales
[S3-vlan10]quit
[S3]vlan 20
[S3-vlan20]description Finances
[S3-vlan20]quit
[S3]interface Ethernet 0/0/1
[S3-Ethernet0/0/1]port link-type access
[S3-Ethernet0/0/1]port default vlan 10
[S3-Ethernet0/0/1]quit
[S3]interface Ethernet 0/0/2
[S3-Ethernet0/0/2]port link-type access
[S3-Ethernet0/0/2]port default vlan 20
[S3-Ethernet0/0/2]quit
[S3]interface GigabitEthernet 0/0/1
[S3-GigabitEthernet0/0/1]port link-type trunk
[S3-GigabitEthernet0/0/1]port trunk allow-pass vlan 10 20
[S3-GigabitEthernet0/0/1]quit
[S3]interface GigabitEthernet 0/0/2
[S3-GigabitEthernet0/0/2]port link-type trunk
```

```
[S3-GigabitEthernet0/0/2]port trunk allow-pass vlan 10 20
```

(4) 对交换机 S4 进行基本配置。

```
<Huawei>system-view
[Huawei]sysname S4
[S4]vlan 10
[S4-vlan10]description Sales
[S4-vlan10]quit
[S4]vlan 20
[S4-vlan20]description Finances
[S4-vlan20]quit
[S4]interface Ethernet 0/0/1
[S4-Ethernet0/0/1]port link-type access
[S4-Ethernet0/0/1]port default vlan 10
[S4-Ethernet0/0/1]quit
[S4]interface Ethernet 0/0/2
[S4-Ethernet0/0/2]port link-type access
[S4-Ethernet0/0/2]port default vlan 20
[S4-Ethernet0/0/2]quit
[S4]interface GigabitEthernet 0/0/1
[S4-GigabitEthernet0/0/1]port link-type trunk
[S4-GigabitEthernet0/0/1]port trunk allow-pass vlan 10 20
[S4-GigabitEthernet0/0/1]quit
[S4]interface GigabitEthernet 0/0/2
[S4-GigabitEthernet0/0/2]port link-type trunk
[S4-GigabitEthernet0/0/2]port trunk allow-pass vlan 10 20
```

步骤 3：在交换机上配置 MSTP。

(1) 在交换机 S1 上配置 MSTP。

```
[S1]stp mode mstp                                  //设置模块为 MSTP
[S1]stp region-configuration                       //进入设备的 MST 域视图
[S1-mst-region]region-name v1020                   //将 MST 域名修改为"v1020"
[S1-mst-region]revision-level 1                    //将 MST 域的修订级别修改为 1
[S1-mst-region]instance 1 vlan 10                  //创建实例 1，将 VLAN10 映射到实例 1
[S1-mst-region]instance 2 vlan 20                  //创建实例 2，将 VLAN20 映射到实例 2
[S1-mst-region]active region-configuration         //激活域
[S1-mst-region]quit
[S1]stp instance 0 root primary     //除了 VLAN10 和 VLAN20，其他 VLAN 保持在默认的实例 0 内
[S1]stp instance 1 root primary                    //将 S1 指定为实例 1 的主根桥
[S1]stp instance 2 root secondary                  //将 S1 指定为实例 2 的次根桥
[S1]stp enable                                     //开启 STP
```

(2) 在交换机 S2 上配置 MSTP。

```
[S2]stp mode mstp                                  //设置模块为 MSTP
[S2]stp region-configuration                       //进入设备的 MST 域视图
```

```
[S2-mst-region]region-name v1020              //将MST域名修改为"v1020"
[S2-mst-region]revision-level 1               //将MST域的修订级别修改为1
[S2-mst-region]instance 1 vlan 10             //创建实例1,将VLAN10映射到实例1
[S2-mst-region]instance 2 vlan 20             //创建实例2,将VLAN20映射到实例2
[S2-mst-region]active region-configuration    //激活域
[S2-mst-region]quit
[S2]stp instance 0 priority 0          //除了VLAN10和VLAN20,其他VLAN保持在默认的实例0内
[S2]stp instance 1 priority 4096              //将S2指定为实例1的次根桥
[S2]stp instance 2 priority 0                 //将S2指定为实例2的主根桥
[S2]stp enable                                //开启STP
```

(3) 在交换机 S3 上配置 MSTP。

```
[S3]stp mode mstp                             //配置模块为MSTP
[S3]stp region-configuration                  //进入设备的MST域视图
[S3-mst-region]region-name v1020              //将MST域名修改为"v1020"
[S3-mst-region]revision-level 1               //将MST域的修订级别修改为1
[S3-mst-region]instance 1 vlan 10             //创建实例1,将VLAN10映射到实例1
[S-mst-region]instance 2 vlan 20              //创建实例2,将VLAN20映射到实例2
[S3-mst-region]active region-configuration    //激活域
[S3-mst-region]quit
[S3]stp enable                                //开启STP
```

(4) 在交换机 S4 上配置 MSTP。

```
[S4]stp mode mstp                             //配置模块为MSTP
[S4]stp region-configuration                  //进入设备的MST域视图
[S4-mst-region]region-name v1020              //将MST域名修改为"v1020"
[S4-mst-region]revision-level 1               //将MST域的修订级别修改为1
[S4-mst-region]instance 1 vlan 10             //创建实例1,将VLAN10映射到实例1
[S4-mst-region]instance 2 vlan 20             //创建实例2,将VLAN20映射到实例2
[S4-mst-region]active region-configuration    //激活域
[S4-mst-region]quit
[S4]stp enable                                //开启STP
```

小贴士

在以上配置中,"stp instance 0 root primary"命令等同于"stp instance 0 priority 0"命令。优先级的取值范围是 0~65 535,且为 4 096 的倍数,如 4 096、8 192 等,默认为 32 768。

任务验收

(1) 使用 "display vlan" 命令验证各交换机上的 VLAN 配置信息。

(2) 使用 "display stp" 命令查看交换机的 STP 状态,以交换机 S1 为例。

```
[S1]display stp
```

```
-------[CIST Global Info][Mode MSTP]-------
CIST Bridge              :0         .4c1f-cc60-485e
Config Times             :Hello 2s MaxAge 20s FwDly 15s MaxHop 20
Active Times             :Hello 2s MaxAge 20s FwDly 15s MaxHop 20
CIST Root/ERPC           :0         .4c1f-cc60-485e / 0
CIST RegRoot/IRPC        :0         .4c1f-cc60-485e / 0
......                                           //此处省略部分内容
//这里可以看到"Mode MSTP"表示的是MSTP模式
```

（3）使用"display stp region-configuration"命令查看当前生效的 MST 域配置信息，以交换机 S1 为例。

```
[S1]display stp region-configuration
Oper configuration
  Format selector     :0                   //格式选择器
  Region name         :v1020               //配置名称
  Revision level      :1                   //修订级别

  Instance   VLANs Mapped                  //实例和VLAN映射集
     0       1 to 9, 11 to 19, 21 to 4094
     1       10
     2       20
```

（4）使用"display stp brief"命令查看交换机的 STP 状态，以交换机 S3 为例。

```
[S3]display stp brief
MSTID  Port                    Role    STP State    Protection
  0    Ethernet0/0/1           DESI    FORWARDING   NONE
  0    Ethernet0/0/2           DESI    FORWARDING   NONE
  0    GigabitEthernet0/0/1    ROOT    FORWARDING   NONE
  0    GigabitEthernet0/0/2    ALTE    DISCARDING   NONE
  1    Ethernet0/0/1           DESI    FORWARDING   NONE
  1    GigabitEthernet0/0/1    ROOT    FORWARDING   NONE
  1    GigabitEthernet0/0/2    ALTE    DISCARDING   NONE
  2    Ethernet0/0/2           DESI    FORWARDING   NONE
  2    GigabitEthernet0/0/1    ALTE    DISCARDING   NONE
  2    GigabitEthernet0/0/2    ROOT    FORWARDING   NONE
//可以看到，在MST1中，S3的GE 0/0/2端口的角色为替代端口，而且状态为丢弃；在MST2中，S3的GE
0/0/1端口的角色为替代端口，而且状态为丢弃，这是符合我们预期的
```

（5）在 PC1 上使用"ping"命令测试 PC1 与 PC3 的连通性。可以发现刚开始 PC1 与 PC3 是连通的，将交换机 S3 的 GE 0/0/1 端口关闭后，PC1 与 PC3 会断开 10 秒左右，之后又连通了，连通性测试结果如图 10.2 所示。

（6）使用"display stp brief"命令观察交换机 S3 其他端口的角色及状态的变化（在交换机 S3 的 GE 0/0/1 端口关闭后）。

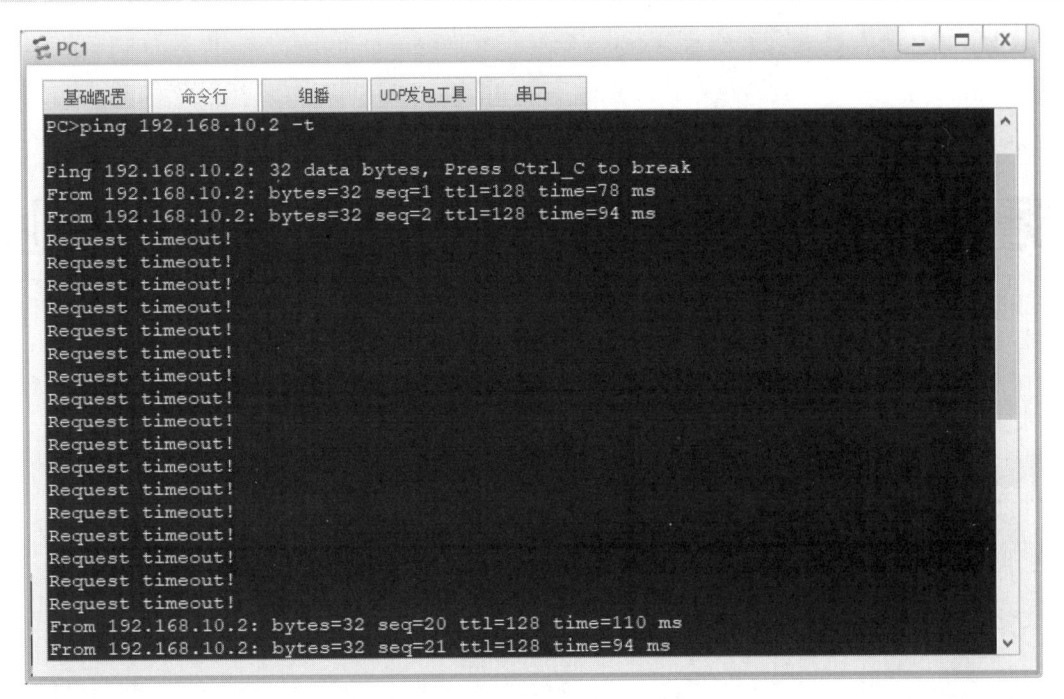

```
┌ PC1                                                    _ □ X
  基础配置   命令行   组播   UDP发包工具   串口
  PC>ping 192.168.10.2 -t

  Ping 192.168.10.2: 32 data bytes, Press Ctrl_C to break
  From 192.168.10.2: bytes=32 seq=1 ttl=128 time=78 ms
  From 192.168.10.2: bytes=32 seq=2 ttl=128 time=94 ms
  Request timeout!
  Request timeout!
  Request timeout!
  Request timeout!
  Request timeout!
  Request timeout!
  Request timeout!
  Request timeout!
  Request timeout!
  Request timeout!
  Request timeout!
  Request timeout!
  Request timeout!
  From 192.168.10.2: bytes=32 seq=20 ttl=128 time=110 ms
  From 192.168.10.2: bytes=32 seq=21 ttl=128 time=94 ms
```

图 10.2 连通性测试结果

```
[S3]interface GigabitEthernet 0/0/1
[S3-GigabitEthernet0/0/1]shutdown
[S3]display stp brief
 MSTID    Port                      Role    STP State    Protection
   0      Ethernet0/0/1             DESI    FORWARDING   NONE
   0      Ethernet0/0/2             DESI    FORWARDING   NONE
   0      GigabitEthernet0/0/2      ROOT    FORWARDING   NONE
   1      Ethernet0/0/1             DESI    FORWARDING   NONE
   1      GigabitEthernet0/0/2      ROOT    FORWARDING   NONE
   2      Ethernet0/0/2             DESI    FORWARDING   NONE
   2      GigabitEthernet0/0/2      ROOT    FORWARDING   NONE
//可以发现，当拓扑结构发生变化时，GE 0/0/2 端口从 Discarding 状态进入 Forwarding 状态
```

任务 11　实现部门计算机动态获取 IP 地址

任务目标

1. 掌握 DHCP 的原理和作用。
2. 掌握配置交换机的 DHCP 技术。

任务描述

某公司的员工反映经常出现 IP 地址冲突的情况，影响上网。网络管理员小赵决定在整个局域网中统一规划 IP 地址，使用户以动态获取地址的方式接入局域网，既节约地址空间，又避免地址冲突现象的发生。

任务要求

在企业网络中，DHCP（Dynamic Host Configuration Protocol，动态主机配置协议）技术可以有规划地分配 IP 地址，避免因用户私设 IP 地址引起地址冲突。三层交换机可以提供 DHCP 服务功能，能够为用户动态地分配 IP 地址、推送 DNS 服务地址等网络参数，从而实现用户"零配置"上网。

（1）为了实现部门计算机动态获取 IP 地址，网络拓扑结构如图 11.1 所示。

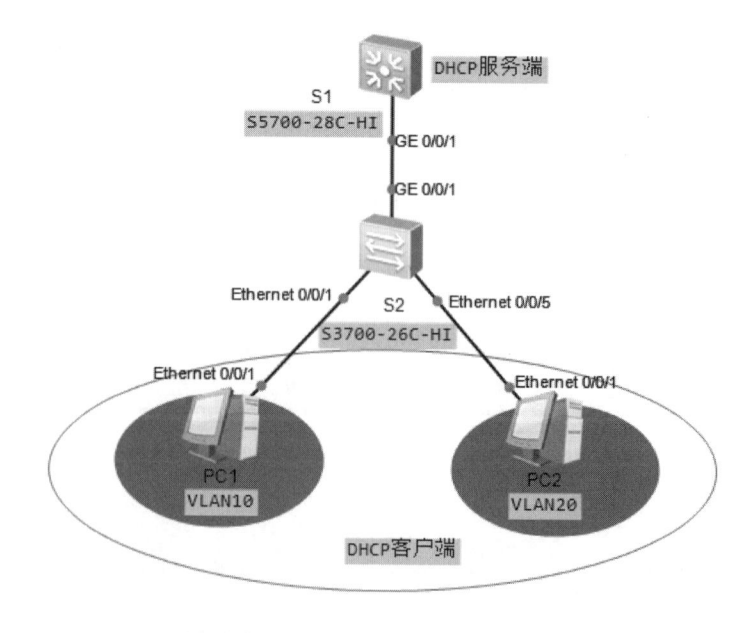

图 11.1　实现部门计算机动态获取 IP 地址的网络拓扑结构

（2）在交换机 S2 上划分两个 VLAN（VLAN10、VLAN20），并将 GE 0/0/1 端口模式设置为 Trunk，划分情况如表 11.1 所示。

表 11.1　二层交换机 S2 的 VLAN 划分情况

VLAN 编号	端 口 范 围	端 口 模 式
10	Ethernet 0/0/1～Ethernet 0/0/4	Access
20	Ethernet 0/0/5～Ethernet 0/0/8	Access
—	GE 0/0/1	Trunk

（3）在交换机 S1 上划分两个 VLAN（VLAN10、VLAN20），配置 VLANIF 的 IP 地址，并将 GE 0/0/1 端口模式设置为 Trunk，划分情况如表 11.2 所示。

表 11.2　三层交换机 S1 的 VLAN 划分情况

VLAN 编号	IP 地址
10	192.168.10.254/24
20	192.168.20.254/24

（4）PC1、PC2 的端口及 IP 地址设置如表 11.3 所示。

表 11.3　PC1、PC2 的端口及 IP 地址设置

设 备 名 称	本 端 端 口	IP 地址/子网掩码	网　关	对端设备与端口
PC1	Ethernet 0/0/1	动态获取	—	S2:Ethernet 0/0/1
PC2	Ethernet 0/0/1	动态获取	—	S2:Ethernet 0/0/5

（5）通过在交换机 S1 上划分两个 VLAN，同时开启 DHCP 服务，可以使连接在交换机上的不同 VLAN 的计算机获得相应的 IP 地址，对 192.168.10.0/24 网段保留前 53 个 IP 地址作为备用，对 192.168.20.0/24 网段保留前 100 个 IP 地址作为备用，最终实现全网互通。

任务实施

步骤 1：按照图 11.1 搭建网络拓扑，连线全部使用直通线，开启所有设备电源，并为每台计算机设置相应的 IP 地址和子网掩码。

步骤 2：对交换机进行基本配置。

（1）配置二层交换机的名称为"S2"，在交换机上划分两个 VLAN，分别为 VLAN10 和 VLAN20，并按要求为两个 VLAN 分配端口，具体命令如下。

```
<Huawei>system-view
[Huawei]sysname S2
[S2]vlan batch 10 20
[S2]port-group 1
[S2-port-group-1]group-member Ethernet 0/0/1 to Ethernet 0/0/4
[S2-port-group-1]port link-type access
```

```
[S2-Ethernet0/0/1]port link-type access
[S2-Ethernet0/0/2]port link-type access
[S2-Ethernet0/0/3]port link-type access
[S2-Ethernet0/0/4]port link-type access
[S2-port-group-1]port default vlan 10
[S2-Ethernet0/0/1]port default vlan 10
[S2-Ethernet0/0/2]port default vlan 10
[S2-Ethernet0/0/3]port default vlan 10
[S2-Ethernet0/0/4]port default vlan 10
[S2-port-group-1]quit
[S2]port-group 2
[S2-port-group-2]group-member Ethernet 0/0/5 to Ethernet 0/0/8
[S2-port-group-2]port link-type access
[S2-Ethernet0/0/5]port link-type access
[S2-Ethernet0/0/6]port link-type access
[S2-Ethernet0/0/7]port link-type access
[S2-Ethernet0/0/8]port link-type access
[S2-port-group-2]port default vlan 20
[S2-Ethernet0/0/5]port default vlan 20
[S2-Ethernet0/0/6]port default vlan 20
[S2-Ethernet0/0/7]port default vlan 20
[S2-Ethernet0/0/8]port default vlan 20
[S2-port-group-2]quit
```

（2）配置三层交换机的名称为"S1"，在交换机上划分两个VLAN，分别是VLAN10和VLAN20，具体命令如下。

```
<Huawei>system-view
[Huawei]sysname S1
[S1]vlan batch 10 20
```

步骤3：配置交换机端口模式为Trunk，并允许VLAN10和VLAN20通过。

（1）配置二层交换机S2的GE 0/0/1端口，具体命令如下。

```
[S2]interface GigabitEthernet 0/0/1
[S2-GigabitEthernet0/0/1]port link-type trunk
[S2-GigabitEthernet0/0/1]port trunk allow-pass vlan 10 20
```

（2）配置三层交换机S1的GE 0/0/1端口，具体命令如下。

```
[S1]interface GigabitEthernet 0/0/1
[S1-GigabitEthernet0/0/1]port link-type trunk
[S1-GigabitEthernet0/0/1]port trunk allow-pass vlan 10 20
```

步骤4：开启交换机的DHCP功能，具体命令如下。

```
[S1]dhcp enable
```

步骤5：配置交换机的DHCP服务，具体命令如下。

```
[S1]ip pool vlan10                                          //创建地址池,名称为VLAN10
[S1-ip-pool-vlan10]network 192.168.10.0 mask 255.255.255.0  //配置可分配的网段范围
[S1-ip-pool-vlan10]gateway-list 192.168.10.254              //配置出口网关地址
[S1-ip-pool-vlan10]lease 5                                  //租期为5天
[S1-ip-pool-vlan10]dns-list 114.114.114.114                 //配置DNS服务器地址
[S1-ip-pool-vlan10]quit
[S1]ip pool vlan20
[S1-ip-pool-vlan20]network 192.168.20.0 mask 255.255.255.0
[S1-ip-pool-vlan20]gateway-list 192.168.20.254
[S1-ip-pool-vlan20]lease 5
[S1-ip-pool-vlan20]dns-list 8.8.8.8
[S1-ip-pool-vlan20]quit
```

步骤6：在配置的交换机 S1 上划分每个 VLAN 的 VLANIF 端口的 IP 地址，同时开启 VLAN 的 VLANIF 端口的 DHCP 功能，具体命令如下。

```
[S1]interface Vlanif 10
[S1-Vlanif10]ip add 192.168.10.254 24
[S1-Vlanif10]dhcp select global                             //配置设备指定端口采取全局地址
[S1-Vlanif10]quit
[S1]interface Vlanif 20
[S1-Vlanif20]ip add 192.168.20.254 24
[S1-Vlanif20]dhcp select global
[S1-Vlanif20]quit
```

步骤7：设置保留的 IP 地址。

在配置 DHCP 服务时，通常需要保留部分 IP 地址，以固定分配的方式给服务器或其他网络设备使用。例如，在本任务中，交换机两个 VLAN 端口的 IP 地址属于固定分配，这些作为保留的 IP 地址就不能以 DHCP 方式再分配给其他计算机了。

在本任务中，要对 192.168.10.0/24 网段保留前 53 个 IP 地址作为备用，对 192.168.20.0/24 网段保留前 100 个 IP 地址作为备用，具体实现命令如下。

```
[S1]ip pool vlan10
[S1-ip-pool-vlan10]excluded-ip-address 192.168.10.201 192.168.10.253
[S1-ip-pool-vlan10]quit
[S1]ip pool vlan20
[S1-ip-pool-vlan20]excluded-ip-address 192.168.20.154 192.168.20.253
```

任务验收

（1）设置计算机以 DHCP 方式获取 IP 地址。右击 PC1，在弹出的快捷菜单中选择"设置"命令，打开 PC1 的设置对话框。首先在"基础配置"选项卡的"IPv4 配置"选区中，选中"DHCP"单选按钮，然后单击对话框右下角的"应用"按钮，如图 11.2 所示。

（2）选择 PC1 的设置对话框中的"命令行"选项卡，在其中输入并执行"ipconfig"命令，查看 PC1 的 IP 地址，如图 11.3 所示。

图 11.2　PC1 的设置对话框

图 11.3　查看 PC1 的 IP 地址

（3）使用同样的方法，为另一台计算机设置 DHCP 获取 IP 地址的方式，并查看计算机获取的 IP 地址信息，最后得到如表 11.4 所示的内容。

表 11.4　计算机获取的 IP 地址信息

计算机	IP 地址	子网掩码	网关	DNS 地址
PC1	192.168.10.200	255.255.255.0	192.168.10.254	114.114.114.114
PC2	192.168.20.153	255.255.255.0	192.168.20.254	8.8.8.8

（4）使用"ping"命令测试其他计算机的连通情况。可以得出结论，当前网络中的计算机之间是连通的，连通性测试结果如图 11.4 所示。

图 11.4　连通性测试结果

任务 12　提高网络的稳定性

任务目标

1．掌握 VRRP 的原理和作用。
2．掌握配置交换机的 VRRP 服务。

任务描述

某公司北京总部的网络承担了连接全国各地分公司网络的任务。该总部网络中心采用多台万兆交换机，内部网络按照业务规划有 2 个部门 VLAN。为了增强总部核心网络的稳定性，要求网络管理员小赵在三层网络上配置 VRRP（Virtual Router Redundancy Protocol，虚拟路由器冗余协议），实现网关冗余，为用户提供透明的切换功能，从而提高网络的稳定性。

任务要求

（1）为了提高网络的稳定性，网络拓扑结构如图 12.1 所示。

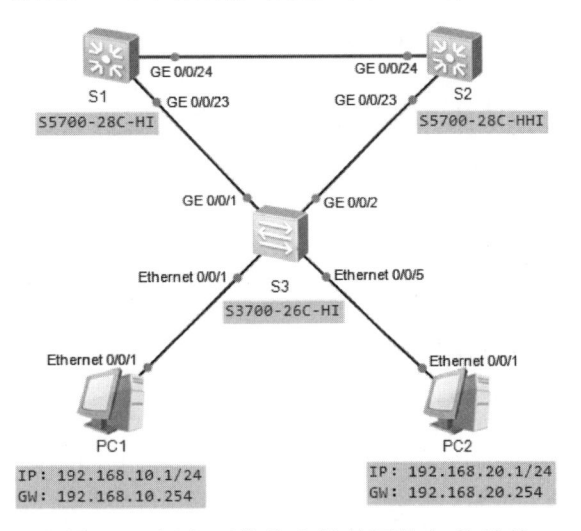

图 12.1　提高网络稳定性的网络拓扑结构

（2）在交换机 S1、S2 和 S3 上划分 VLAN10 和 VLAN20，端口范围、IP 地址和端口模式等情况如表 12.1 所示。

表 12.1　交换机的 VLAN 划分情况

设备名称	VLAN 编号	端口范围	IP 地址/端口模式
S1	10	—	192.168.10.100/24
	20	—	192.168.20.100/24
	—	GE 0/0/23	Trunk
		GE 0/0/24	Trunk
S2	10	—	192.168.10.200/24
	20	—	192.168.20.200/24
	—	GE 0/0/23	Trunk
		GE 0/0/24	Trunk
S3	10	Ethernet 0/0/1～Ethernet 0/0/4	Access
	20	Ethernet 0/0/5～Ethernet 0/0/8	Access
	—	GE 0/0/1	Trunk
		GE 0/0/2	Trunk

（3）PC1、PC2 的端口及 IP 地址设置如表 12.2 所示。

表 12.2　PC1、PC2 的端口及 IP 地址设置

设备名称	本端端口	IP 地址/子网掩码	所属 VLAN	对端设备与端口
PC1	Ethernet 0/0/1	192.168.10.1/24	10	S3:Ethernet 0/0/1
PC2	Ethernet 0/0/1	192.168.20.1/24	20	S3:Ethernet 0/0/5

（4）在交换机 S1、S2 上配置 VRRP 服务，使连接在二层交换机上的不同 VLAN 的计算机实现透明的切换，提高网络的稳定性。

任务实施

步骤 1：按照图 12.1 搭建网络拓扑，连线全部使用直通线，开启所有设备电源，并为每台计算机设置相应的 IP 地址和子网掩码。

步骤 2：对交换机进行基本配置。

（1）配置二层交换机的名称为"S3"，在交换机上划分两个 VLAN，分别为 VLAN10 和 VLAN20，并按要求为两个 VLAN 分配端口。

```
<Huawei>system-view
[Huawei]sysname S3
[S3]vlan batch 10 20
[S3]port-group 1
[S3-port-group-1]group-member Ethernet 0/0/1 to Ethernet 0/0/4
[S3-port-group-1]port link-type access
```

```
[S3-Ethernet0/0/1]port link-type access
[S3-Ethernet0/0/2]port link-type access
[S3-Ethernet0/0/3]port link-type access
[S3-Ethernet0/0/4]port link-type access
[S3-port-group-1]port default vlan 10
[S3-Ethernet0/0/1]port default vlan 10
[S3-Ethernet0/0/2]port default vlan 10
[S3-Ethernet0/0/3]port default vlan 10
[S3-Ethernet0/0/4]port default vlan 10
[S3-port-group-1]quit
[S3]port-group 2
[S3-port-group-2]group-member Ethernet 0/0/5 to Ethernet 0/0/8
[S3-port-group-2]port link-type access
[S3-Ethernet0/0/5]port link-type access
[S3-Ethernet0/0/6]port link-type access
[S3-Ethernet0/0/7]port link-type access
[S3-Ethernet0/0/8]port link-type access
[S3-port-group-2]port default vlan 20
[S3-Ethernet0/0/5]port default vlan 20
[S3-Ethernet0/0/6]port default vlan 20
[S3-Ethernet0/0/7]port default vlan 20
[S3-Ethernet0/0/8]port default vlan 20
[S3-port-group-2]quit
```

（2）配置三层交换机的名称为"S1"，在交换机上划分两个VLAN，分别为VLAN10和VLAN20。

```
<Huawei>system-view
[Huawei]sysname S1
[S1]vlan batch 10 20                              //划分VLAN10和VLAN20
```

（3）配置三层交换机的名称为"S2"，在交换机上划分两个VLAN，分别为VLAN10和VLAN20。

```
<Huawei>system-view
[Huawei]sysname S2
[S2]vlan batch 10 20                              //划分VLAN10和VLAN20
```

步骤3：配置交换机端口为"Trunk"，并允许VLAN10和VLAN20通过。

（1）配置二层交换机S3的GE 0/0/1和GE 0/0/2端口。

```
[S3]interface GigabitEthernet 0/0/1
[S3-GigabitEthernet0/0/1]port link-type trunk
[S3-GigabitEthernet0/0/1]port trunk allow-pass vlan 10 20
[S3-GigabitEthernet0/0/1]quit
[S3]interface GigabitEthernet 0/0/2
[S3-GigabitEthernet0/0/2]port link-type trunk
```

```
[S3-GigabitEthernet0/0/2]port trunk allow-pass vlan 10 20
[S3-GigabitEthernet0/0/2]quit
```

（2）配置交换机 S1 的 GE 0/0/23 和 GE 0/0/24 端口。

```
[S1]interface GigabitEthernet 0/0/23
[S1-GigabitEthernet0/0/23]port link-type trunk
[S1-GigabitEthernet0/0/23]port trunk allow-pass vlan 10 20
[S1-GigabitEthernet0/0/23]quit
[S1]interface GigabitEthernet 0/0/24
[S1-GigabitEthernet0/0/24]port link-type trunk
[S1-GigabitEthernet0/0/24]port trunk allow-pass vlan 10 20
[S1-GigabitEthernet0/0/24]quit
```

（3）配置交换机 S2 的 GE 0/0/23 和 GE 0/0/24 端口。

```
[S2]interface GigabitEthernet 0/0/23
[S2-GigabitEthernet0/0/23]port link-type trunk
[S2-GigabitEthernet0/0/23]port trunk allow-pass vlan 10 20
[S2-GigabitEthernet0/0/23]quit
[S2]interface GigabitEthernet 0/0/24
[S2-GigabitEthernet0/0/24]port link-type trunk
[S2-GigabitEthernet0/0/24]port trunk allow-pass vlan 10 20
[S2-GigabitEthernet0/0/24]quit
```

步骤 4：配置交换机 VLAN 的 VLANIF 端口的 IP 地址。

（1）配置交换机 S1 上划分的每个 VLAN 的 VLANIF 端口的 IP 地址。

```
[S1]interface Vlanif 10
[S1-Vlanif10]ip add 192.168.10.100 24
[S1-Vlanif10]quit
[S1]interface Vlanif 20
[S1-Vlanif20]ip add 192.168.20.100 24
[S1-Vlanif20]quit
```

（2）配置交换机 S2 上划分的每个 VLAN 的 VLANIF 端口的 IP 地址。

```
[S2]interface Vlanif 10
[S2-Vlanif10]ip add 192.168.10.200 24
[S2-Vlanif10]quit
[S2]interface Vlanif 20
[S2-Vlanif20]ip add 192.168.20.200 24
[S2-Vlanif20]quit
```

步骤 5：配置交换机的 VRRP 服务。

（1）配置交换机 S1 的 VRRP 服务，配置交换机上每个 VLAN 的 VLANIF 端口的 IP 地址、优先级、抢占模式和延迟时间。

```
[S1]interface Vlanif 10
[S1-Vlanif10]vrrp vrid 1 virtual-ip 192.168.10.254        //配置 VLANIF 端口的 IP 地址
```

```
[S1-Vlanif10]vrrp vrid 1 priority 150                    //配置优先级
[S1-Vlanif10]vrrp vrid 1 preempt-mode timer delay 5      //配置抢占模式和延迟时间
[S1-Vlanif10]vrrp vrid 1 track interface GigabitEthernet0/0/23 reduced 50
                                                         //将 GE 0/0/23 配置为跟踪端口
[S1-Vlanif10]quit
[S1]interface Vlanif 20
[S1-Vlanif20]vrrp vrid 2 virtual-ip 192.168.20.254       //配置 VLANIF 端口的 IP 地址
[S1-Vlanif20]vrrp vrid 2 priority 110
[S1-Vlanif20]quit
```

（2）配置交换机 S2 的 VRRP 服务，配置交换机上每个 VLAN 的 VLANIF 端口的 IP 地址、优先级、抢占模式和延迟时间。

```
[S2]interface Vlanif 10
[S2-Vlanif10]vrrp vrid 1 virtual-ip 192.168.10.254
[S2-Vlanif10]vrrp vrid 1 priority 110
[S2-Vlanif10]quit
[S2]interface Vlanif 20
[S2-Vlanif20]vrrp vrid 2 virtual-ip 192.168.20.254
[S2-Vlanif20]vrrp vrid 2 priority 150
[S2-Vlanif20]vrrp vrid 2 preempt-mode timer delay 5
[S1-Vlanif20]vrrp vrid 2 track interface GigabitEthernet0/0/23 reduced 50
[S2-Vlanif20]quit
```

步骤 6：查看交换机的 VRRP 服务。

（1）在交换机 S1 上，使用"display vrrp brief"命令查看当前工作状态。

```
[S1]display vrrp brief
VRID  State      Interface         Type       Virtual IP
--------------------------------------------------------------
1     Master     Vlanif10          Normal     192.168.10.254
2     Backup     Vlanif20          Normal     192.168.20.254
--------------------------------------------------------------
Total:2    Master:1    Backup:1    Non-active:0
```

（2）在交换机 S1 上，使用"display vrrp 1"命令查看当前工作状态。

```
[S1]display vrrp 1
 Vlanif10 | Virtual Router 1
   State : Master
   Virtual IP : 192.168.10.254
   Master IP : 192.168.10.100
   PriorityRun : 150
   PriorityConfig : 150
   MasterPriority : 150
   Preempt : YES   Delay Time : 5 s
```

（3）在交换机 S2 上，使用"display vrrp brief"命令查看当前工作状态。

```
[S2]display vrrp brief
VRID  State      Interface              Type      Virtual IP
--------------------------------------------------------------
1     Backup     Vlanif10               Normal    192.168.10.254
2     Master     Vlanif20               Normal    192.168.20.254
--------------------------------------------------------------
Total:2    Master:1    Backup:1    Non-active:0
```

任务验收

（1）在 PC1 的设置对话框中选择"命令行"选项卡，使用"ping"和"tracert"命令测试 PC1 与 PC2 的连通性，如图 12.2 所示。

（2）断开交换机 S3 右侧 GE 0/0/2 端口的连线，使用"ping"和"tracert"命令再次测试 PC1 与 PC2 的连通性，发现此时有短暂的丢包现象后，又恢复了连通，如图 12.3 所示。由此可以得出结论，当前网络中的所有计算机之间是连通的。

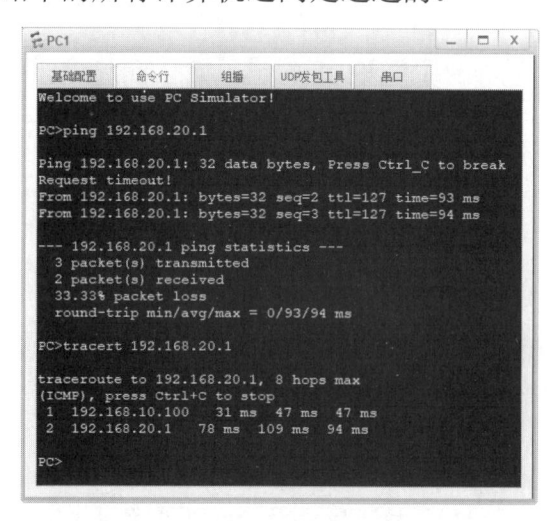

图 12.2　使用"ping"和"tracert"命令测试 PC1 与 PC2 的连通性

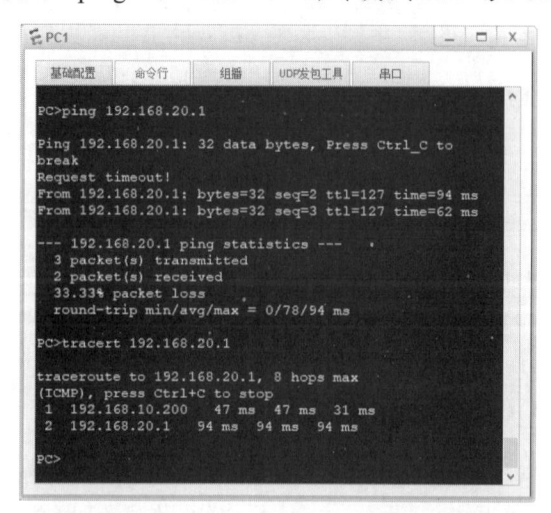

图 12.3　使用"ping"和"tracert"命令再次测试 PC1 与 PC2 的连通性

（3）注意交换机 S1 的状态由 Master 变化为 Backup。

```
[S1]display vrrp brief
VRID  State      Interface          Type       Virtual IP
--------------------------------------------------------------
1     Backup     Vlanif10           Normal     192.168.10.254
2     Backup     Vlanif20           Normal     192.168.20.254
--------------------------------------------------------------
Total:2    Master:0    Backup:2    Non-active:0
```

（4）在交换机 S2 上，使用"display vrrp 1"和"display vrrp brief"命令查看当前工作状态，并注意查看 VRRP 的状态变化。

任务 13 路由器的基本配置

任务目标

1. 理解路由器的工作原理。
2. 熟悉路由器的基本配置。
3. 掌握使用路由器的基本配置命令。

任务描述

某公司因业务发展需求，购买了一批华为路由器来扩展现有的网络，按照公司网络管理的要求，网络管理员小赵在通过路由器的 Console 口连接路由器后，需要完成路由器的配置、管理任务，以及优化网络环境。

任务要求

本次任务包括实现 CLI 管理路由器设备的配置、几种视图模式的进入与退出、Console 口密码的配置、路由器命名、端口与 IP 地址的配置、日期时钟的配置等。

（1）为了实现路由器的基本配置，网络拓扑结构如图 13.1 所示。

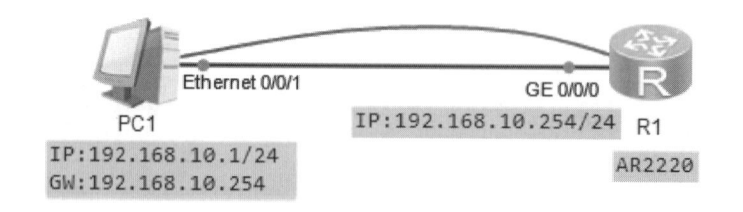

图 13.1 路由器基本配置的网络拓扑结构

（2）路由器 R1 和 PC1 的端口及 IP 地址设置如表 13.1 所示。

表 13.1　路由器 R1 和 PC1 的端口及 IP 地址设置

设 备 名 称	本 端 端 口	IP 地址/子网掩码	网　　关	对端设备与端口
R1	GE 0/0/0	192.168.10.254/24	—	PC1: Ethernet 0/0/1
PC1	Ethernet 0/0/1	192.168.10.1/24	192.168.10.254	R1: GE 0/0/0

任务实施

步骤 1：按照图 13.1 搭建网络拓扑，连线全部使用直通线，开启所有设备电源，并为每台计算机设置相应的 IP 地址和子网掩码。

步骤 2：双击 PC1，在打开的设置对话框中选择"串口"选项卡，在"设置"选区中设置参数，如图 13.2 所示。

步骤 3：设置为默认参数，并在"命令行"选区中输入"<Huawei>"，单击"连接"按钮，如图 13.3 所示。

图 13.2　选择"串口"选项卡 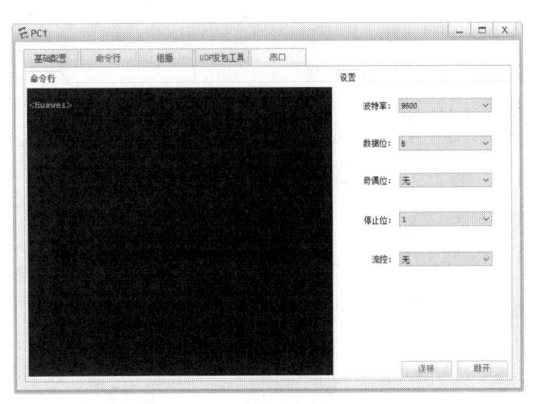　　　　图 13.3　设置参数并连接

步骤 4：用户可以按照相同方式打开路由器设置对话框，还可以对路由器进行必要的设置。在 PC1 上，使用"display version"命令可以查看路由器的软/硬件版本信息，如图 13.4 所示。

图 13.4　查看路由器的软/硬件版本信息

步骤 5：路由器的配置模式切换。

```
<Huawei>                                        //用户视图
<Huawei>system-view                             //进入系统视图
[Huawei]interface GigabitEthernet 0/0/0         //进入端口视图
[Huawei-GigabitEthernet0/0/0]quit               //回到系统视图
[Huawei]quit                                    //回到用户视图
<Huawei>save                                    //保存配置
```

步骤 6：配置路由器的名称。

```
<Huawei>system-view                             //进入系统视图
[Huawei]sysname R1                              //将路由器命名为"R1"
```

步骤 7：配置路由器的时间。

```
<R1>clock timezone BJ add 08:00:00
<R1>clock datetime 12:00:00 2022-12-12
<R1>display clock
2022-12-12 12:00:05
Monday
Time Zone(CST) : UTC+08:00
```

步骤 8：配置设备端口的 IP 地址。

```
<R1>system-view
[R1]int GigabitEthernet 0/0/0                   //进入 GE 0/0/0 端口
[R1-GigabitEthernet0/0/0]ip address 192.168.10.254 255.255.255.0
```

步骤 9：配置路由器的 Console 口密码。

（1）配置登录用户界面的认证方式为密码认证，密码为"Huawei"。

```
[R1]user-interface console 0                                           //进入 Console 口
[R1-ui-console0]authentication-mode password                           //认证方式为密码认证
Please configure the login password (maximum length 16):Huawei         //密码为"Huawei"
[R1-ui-console0]return
<R1>quit
测试：
Login authentication
Password:                                       //此处输入密码"Huawei"，可以进入用户视图
<R1>
```

（2）配置登录用户界面的认证方式为 AAA 认证，用户名为"admin"，密码为"Huawei"。

```
<R1>system-view
[R1]user-interface console 0
[R1-ui-console0]authentication-mode aaa         //认证方式为 AAA 认证
[R1-ui-console0]quit
[R1]aaa
```

```
[R1-aaa]local-user admin password cipher Huawei    //密码为"Huawei"
[R1-aaa]local-user admin service-type terminal     //用户名为"admin",服务类型为终端
[R1-aaa]return
<R1>quit
Username:admin          //此处输入的用户名为"admin",密码为"Huawei",可以进入用户视图
Password:
<R1>
```

任务验收

(1) 使用"display current-configuration"命令查看当前配置。查看交换机的管理方式是否配置成功。

(2) 测试交换机的 Console 口的密码是否已经生效。

(3) 在配置完成后,再次查看网络拓扑结构,可以发现链路中的红色标记已经变成了绿色。这时可以为 PC1 划分一个同网段的 IP 地址(如 192.168.10.1),设置网关为路由器 GE 0/0/0 端口的 IP 地址,并测试它们之间的连通性。

任务 14　利用单臂路由实现部门计算机之间的网络互访

任务目标

1. 了解单臂路由的工作原理。
2. 实现路由器的单臂路由配置。

任务描述

某公司的网络管理员小赵为部门计算机划分了 VLAN 后，发现两个部门的计算机之间无法通信，但有时两个部门的计算机需要进行通信，因此网络管理员需要通过简单的方法来实现通信功能。利用路由器的单臂路由可以解决这个问题。

任务要求

（1）利用单臂路由实现部门计算机之间的网络互访，网络拓扑结构如图 14.1 所示。

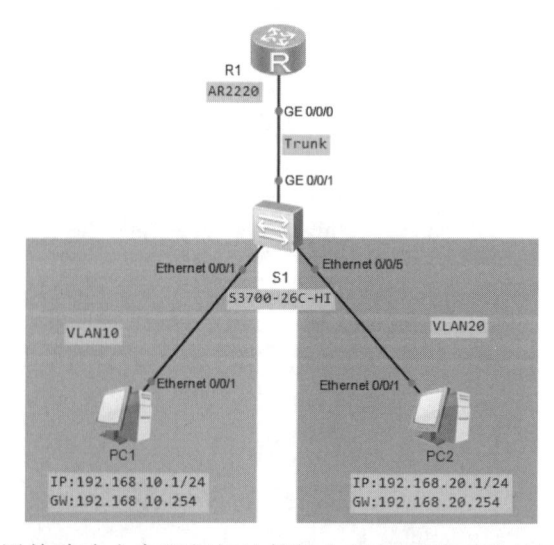

图 14.1　利用单臂路由实现部门计算机之间的网络互访的网络拓扑结构

（2）交换机 S1 的 VLAN 划分情况如表 14.1 所示。

表 14.1　交换机 S1 的 VLAN 划分情况

VLAN 编号	端　口　范　围	连接的计算机
10	Ethernet 0/0/1	PC1
20	Ethernet 0/0/5	PC2

（3）路由器 R1 和 PC1、PC2 的端口及 IP 地址设置如表 14.2 所示。

表 14.2　路由器 R1 和 PC1、PC2 的端口及 IP 地址设置

设 备 名 称	端　　　口	IP 地址/子网掩码	默 认 网 关
R1	GE 0/0/0.1	192.168.10.254/24	—
	GE 0/0/0.2	192.168.20.254/24	—
PC1	Ethernet 0/0/1	192.168.10.1/24	192.168.10.254
PC2	Ethernet 0/0/1	192.168.20.1/24	192.168.20.254

（4）在路由器上配置单臂路由，从而实现两台计算机正常通信。

任务实施

步骤 1：按照图 14.1 搭建网络拓扑，连线全部使用直通线，开启所有设备电源，并为每台计算机设置相应的 IP 地址和子网掩码。

步骤 2：对交换机 S1 进行基本配置。

```
<Huawei>system-view
[Huawei]sysname S1
[S1]vlan batch 10 20
[S1]interface Ethernet 0/0/1
[S1-Ethernet0/0/1]port link-type access
[S1-Ethernet0/0/1]port default vlan 10
[S1-Ethernet0/0/1]quit
[S1]interface Ethernet 0/0/5
[S1-Ethernet0/0/5]port link-type access
[S1-Ethernet0/0/5]port default vlan 20
[S1-Ethernet0/0/5]quit
[S1]interface GigabitEthernet 0/0/1
[S1-GigabitEthernet0/0/1]port link-type trunk
[S1-GigabitEthernet0/0/1]port trunk allow-pass vlan 10 20
[S1-GigabitEthernet0/0/1]quit
```

步骤 3：对路由器 R1 进行基本配置。

```
<Huawei>system-view
[Huawei]sysname R1
[R1]interface GigabitEthernet 0/0/0.1                    //进入 GE 0/0/0.1 子端口
```

```
[R1-GigabitEthernet0/0/0.1]ip add 192.168.10.254 24    //设置 IP 地址
[R1-GigabitEthernet0/0/0.1]dot1q termination vid 10    //封装 802.1q 协议
[R1-GigabitEthernet0/0/0.1]arp broadcast enable        //开启 ARP 广播功能
[R1-GigabitEthernet0/0/0.1]quit
[R1]int g0/0/0.2                                       //进入 GE 0/0/0.2 子端口
[R1-GigabitEthernet0/0/0.2]ip add 192.168.20.254 24    //设置 IP 地址
[R1-GigabitEthernet0/0/0.2]dot1q termination vid 20    //封装 802.1q 协议
[R1-GigabitEthernet0/0/0.2]arp broadcast enable        //开启 ARP 广播功能
[R1-GigabitEthernet0/0/0.2]quit
```

步骤 4：在路由器 R1 上查看端口状态。

```
[R1]display ip interface brief
*down: administratively down
……                                        //此处省略部分内容
Interface                   IP Address/Mask      Physical    Protocol
GigabitEthernet0/0/0        unassigned           up          down
GigabitEthernet0/0/0.1      192.168.10.254/24    up          up
GigabitEthernet0/0/0.2      192.168.20.254/24    up          up
GigabitEthernet0/0/1        unassigned           down        down
GigabitEthernet0/0/2        unassigned           down        down
NULL0                       unassigned           up          up(s)
//可以观察到，两个子端口的物理状态和协议状态都正常
```

任务验收

（1）使用"display ip routing-table"命令查看路由器 R1 的路由表。观察路由器 R1 的路由表中是否已经有 192.168.10.1/24 和 192.168.20.1/24 路由条目。

（2）PC1 和 PC2 分别属于 VLAN10 和 VLAN20，交换机 S1 是一个二层交换机，为实现 VLAN10 和 VLAN20 中的计算机相互通信，要增加一个路由器来转发 VLAN 之间的数据包，路由器与交换机之间使用单条链路相连，这条链路又称主干（Trunk），所有数据包的进出都要通过路由器 R1 的 GE 0/0/0 端口实现。

在配置完成后，再次查看网络拓扑结构，可以发现链路中的红色标记已经变成了绿色。这时可以使用"ping"命令测试 PC1 与 PC2 的连通性，连通性测试结果如图 14.2 所示，可以发现它们之间是连通的。这说明路由器的单臂路由功能发挥了作用。

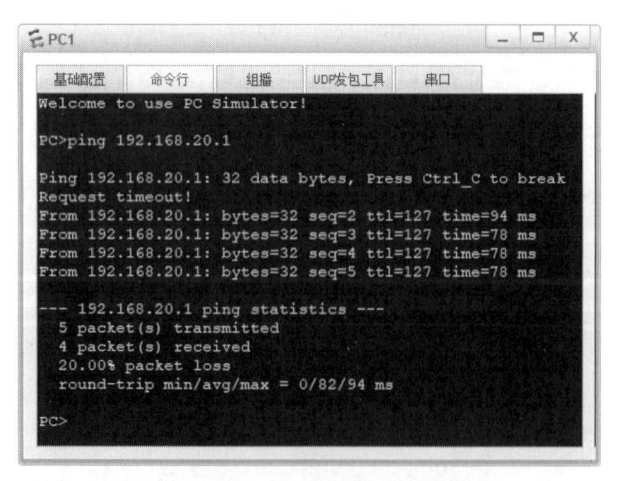

图 14.2 连通性测试结果

任务 15 使用静态路由实现网络连通

任务目标

1. 了解路由表的产生方式。
2. 了解静态路由的作用和工作原理。
3. 实现路由器和三层交换机的静态路由协议配置。

任务描述

某公司的规模较小,网络管理员应非常清楚地了解网络的拓扑结构,以便设置正确的路由信息。公司的网络管理员小赵经过考虑,认为静态路由适用于比较简单的网络环境,由于该公司网络规模较小且不经常变动,使用静态路由比较合适。因此,小赵在公司的路由器、交换机与运营商路由器之间使用静态路由,从而实现网络连通。

任务要求

(1) 使用静态路由实现网络连通,网络拓扑结构如图 15.1 所示。

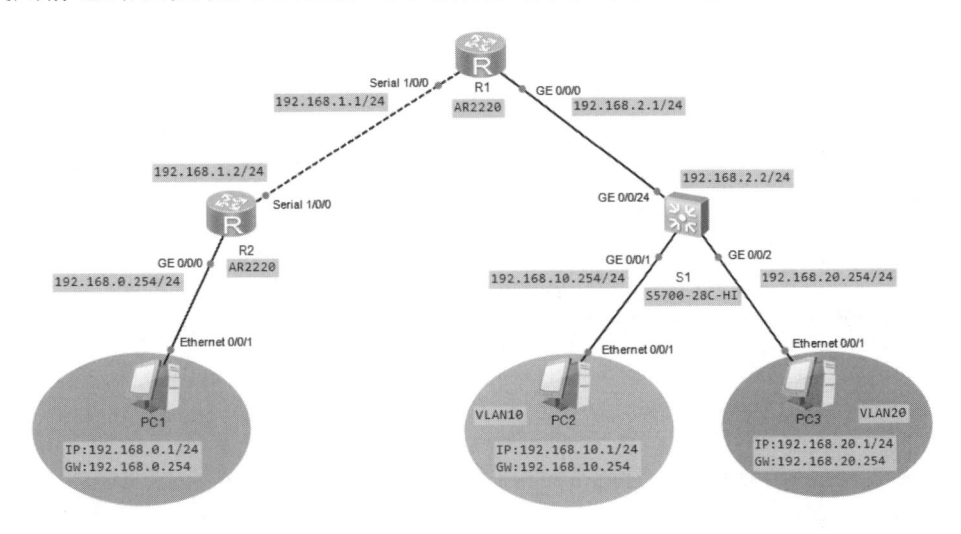

图 15.1 使用静态路由实现网络连通的网络拓扑结构

（2）路由器和交换机的端口及 IP 地址设置如表 15.1 所示。

表 15.1　路由器和交换机的端口及 IP 地址设置

设备名称	端　　口	IP 地址/子网掩码
R1	GE 0/0/0	192.168.2.1/24
	Serial 1/0/0	192.168.1.1/24
R2	Serial 1/0/0	192.168.1.2/24
	GE 0/0/0	192.168.0.254/24
S1	GE 0/0/1（VLANIF10）	192.168.10.254/24
	GE 0/0/2（VLANIF20）	192.168.20.254/24
	GE 0/0/24（VLANIF30）	192.168.2.2/24

（3）PC1~PC3 的端口及 IP 地址设置如表 15.2 所示。

表 15.2　PC1~PC3 的端口及 IP 地址设置

设备名称	本端端口	IP 地址/子网掩码	网　关	对端设备与端口
PC1	Ethernet 0/0/1	192.168.0.1/24	192.168.0.254	R2:GE 0/0/0
PC2	Ethernet 0/0/1	192.168.10.1/24	192.168.10.254	S1:GE 0/0/1
PC3	Ethernet 0/0/1	192.168.20.1/24	192.168.20.254	S1:GE 0/0/2

（4）在路由器和交换机上运行静态路由协议，实现网络连通。

任务实施

步骤 1：按照图 15.1 搭建网络拓扑，在路由器的 Serial 1/0/0 端口位置上添加 2SA 模块，路由器之间的连线使用 Serial 串口线，其他使用直通线，开启所有设备电源，并为每台计算机设置相应的 IP 地址和子网掩码。

步骤 2：设置交换机和路由器的基本配置。

（1）对交换机 S1 进行基本配置。

```
<Huawei>system-view
[Huawei]sysname S1                                    //配置交换机名称
[S1]vlan batch 10 20 30                               //创建 VLAN10、VLAN20 和 VLAN30
[S1]interface GigabitEthernet 0/0/1
[S1-GigabitEthernet0/0/1]port link-type access
[S1-GigabitEthernet0/0/1]port default vlan 10
[S1-GigabitEthernet0/0/1]quit
[S1]interface GigabitEthernet 0/0/2
[S1-GigabitEthernet0/0/2]port link-type access
[S1-GigabitEthernet0/0/2]port default vlan 20
[S1-GigabitEthernet0/0/2]quit
[S1]interface GigabitEthernet 0/0/24
```

```
[S1-GigabitEthernet0/0/24]port link-type access
[S1-GigabitEthernet0/0/24]port default vlan 30
[S1-GigabitEthernet0/0/24]quit
```

（2）在交换机 S1 上创建 VLANIF 端口，在端口视图中配置 IP 地址。

```
[S1]interface Vlanif 10
[S1-Vlanif10]ip add 192.168.10.254 24
[S1-Vlanif10]quit
[S1]interface Vlanif 20
[S1-Vlanif20]ip add 192.168.20.254 24
[S1-Vlanif20]quit
[S1]interface Vlanif 30
[S1-Vlanif30]ip add 192.168.2.2 24
[S1-Vlanif30]quit
```

（3）对路由器 R1 进行基本配置。

```
<Huawei>system-view
[Huawei]sysname R1
[R1]interface GigabitEthernet 0/0/0
[R1-GigabitEthernet0/0/0]ip add 192.168.2.1 24
[R1-GigabitEthernet0/0/0]quit
[R1]interface Serial 1/0/0
[R1-Serial1/0/0]ip add 192.168.1.1 24
[R1-Serial1/0/0]quit
```

（4）对路由器 R2 进行基本配置。

```
<Huawei>system-view
[Huawei]sysname R2
[R2]interface Serial 1/0/0
[R2-Serial1/0/0]ip add 192.168.1.2 24
[R2-Serial1/0/0]quit
[R2]interface GigabitEthernet 0/0/0
[R2-GigabitEthernet0/0/0]ip add 192.168.0.254 24
[R2-GigabitEthernet0/0/0]quit
```

步骤 3：配置静态路由，实现全网互通。

（1）路由器 R1 不能直接到达的网络都要添加静态路由，分别有 192.168.10.0、192.168.20.0 和 192.168.0.0 三个网络，而路由器 R1 到 192.168.10.0 和 192.168.20.0 这两个网络都要通过交换机 S1 的 GE 0/0/24 端口进行转发，到 192.168.0.0 这个网络要通过路由器 R2 的 Serial 1/0/0 端口进行转发，因此要在路由器 R1 上添加的静态路由如下。

```
[R1]ip route-static 192.168.10.0 255.255.255.0 192.168.2.2
[R1]ip route-static 192.168.20.0 255.255.255.0 192.168.2.2
[R1]ip route-static 192.168.0.0 255.255.255.0 192.168.1.2
```

（2）路由器 R2 不能直接到达的网络都要添加静态路由，分别有 192.168.2.0、192.168.10.0 和 192.168.20.0 三个网络，而路由器 R2 连接这三个网络都要通过路由器 R1 的 Serial 1/0/0 端口转发，Serial 1/0/0 端口的 IP 地址即静态路由中的下一跳地址，因此在路由器 R2 上添加的静态路由如下。

```
[R2]ip route-static 192.168.10.0 255.255.255.0 192.168.1.1
[R2]ip route-static 192.168.20.0 255.255.255.0 192.168.1.1
[R2]ip route-static 192.168.2.0 255.255.255.0 192.168.1.1
```

（3）交换机 S1 不能直接到达的网络都要添加静态路由，分别有 192.168.0.0 和 192.168.1.0 两个网络，而交换机 S1 连接这两个网络都要通过路由器 R1 的 GE 0/0/0 端口转发，GE 0/0/0 端口的 IP 地址即静态路由中的下一跳地址，因此要在交换机 S1 上添加的静态路由如下。

```
[S1]ip route-static 192.168.0.0 255.255.255.0 192.168.2.1
[S1]ip route-static 192.168.1.0 255.255.255.0 192.168.2.1
```

任务验收

（1）在路由器 R1 上，使用"display ip routing-table"命令查看路由表。

```
[R1]display ip routing-table
Route Flags: R - relay, D - download to fib
------------------------------------------------------------
Routing Tables: Public
        Destinations : 14      Routes : 14
Destination/Mask        Proto   Pre  Cost  Flags  NextHop          Interface
127.0.0.0/8             Direct  0    0       D    127.0.0.1        InLoopBack0
127.0.0.1/32            Direct  0    0       D    127.0.0.1        InLoopBack0
127.255.255.255/32      Direct  0    0       D    127.0.0.1        InLoopBack0
192.168.0.0/24          Static  60   0       RD   192.168.1.2      Serial1/0/0
192.168.1.0/24          Direct  0    0       D    192.168.1.1      Serial1/0/0
192.168.1.1/32          Direct  0    0       D    127.0.0.1        Serial1/0/0
192.168.1.2/32          Direct  0    0       D    192.168.1.2      Serial1/0/0
192.168.1.255/32        Direct  0    0       D    127.0.0.1        Serial1/0/0
192.168.2.0/24          Direct  0    0       D    192.168.2.1      GigabitEthernet0/0/0
192.168.2.1/32          Direct  0    0       D    127.0.0.1        GigabitEthernet0/0/0
192.168.2.255/32        Direct  0    0       D    127.0.0.1        GigabitEthernet0/0/0
192.168.10.0/24         Static  60   0       RD   192.168.2.2      GigabitEthernet0/0/0
192.168.20.0/24         Static  60   0       RD   192.168.2.2      GigabitEthernet0/0/0
255.255.255.255/32      Direct  0    0       D    127.0.0.1        InLoopBack0
```

（2）使用 PC1 ping PC2 和 PC3 的 IP 地址以测试连通性，连通性测试结果如图 15.2 所示。

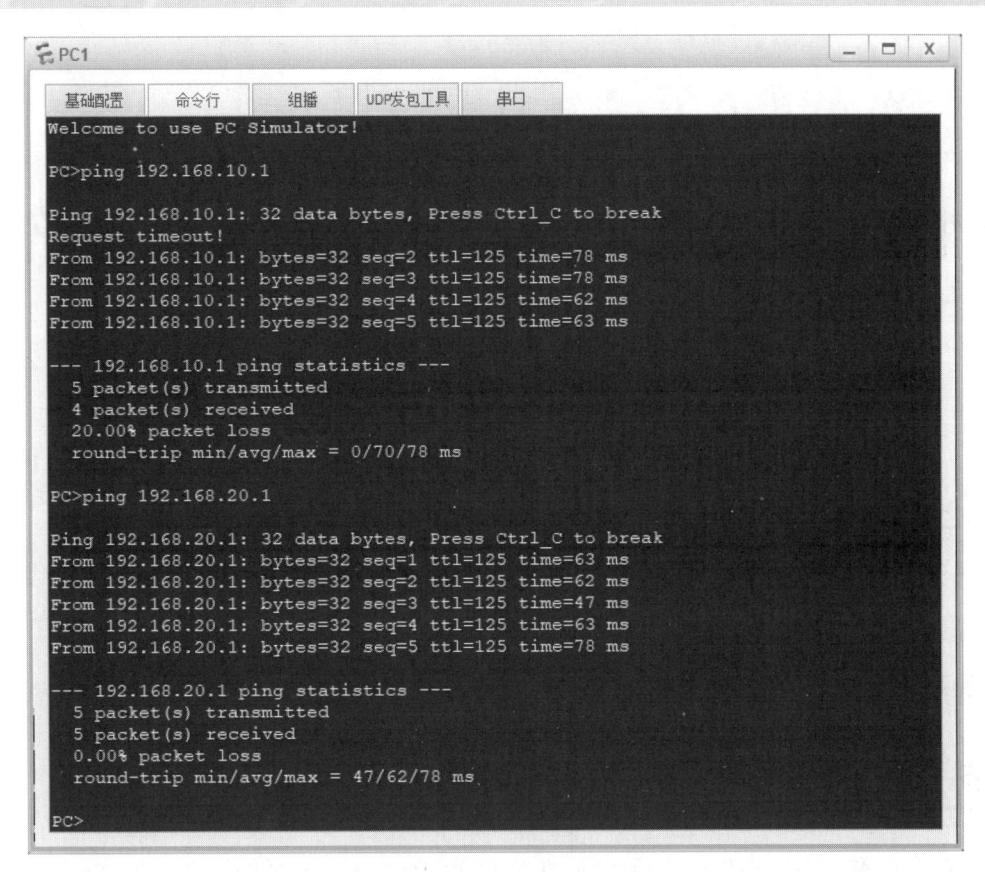

```
PC1                                              _ □ X
基础配置  命令行  组播  UDP发包工具  串口
Welcome to use PC Simulator!
PC>ping 192.168.10.1

Ping 192.168.10.1: 32 data bytes, Press Ctrl_C to break
Request timeout!
From 192.168.10.1: bytes=32 seq=2 ttl=125 time=78 ms
From 192.168.10.1: bytes=32 seq=3 ttl=125 time=78 ms
From 192.168.10.1: bytes=32 seq=4 ttl=125 time=62 ms
From 192.168.10.1: bytes=32 seq=5 ttl=125 time=63 ms

--- 192.168.10.1 ping statistics ---
  5 packet(s) transmitted
  4 packet(s) received
  20.00% packet loss
  round-trip min/avg/max = 0/70/78 ms

PC>ping 192.168.20.1

Ping 192.168.20.1: 32 data bytes, Press Ctrl_C to break
From 192.168.20.1: bytes=32 seq=1 ttl=125 time=63 ms
From 192.168.20.1: bytes=32 seq=2 ttl=125 time=62 ms
From 192.168.20.1: bytes=32 seq=3 ttl=125 time=47 ms
From 192.168.20.1: bytes=32 seq=4 ttl=125 time=63 ms
From 192.168.20.1: bytes=32 seq=5 ttl=125 time=78 ms

--- 192.168.20.1 ping statistics ---
  5 packet(s) transmitted
  5 packet(s) received
  0.00% packet loss
  round-trip min/avg/max = 47/62/78 ms

PC>
```

图 15.2　连通性测试结果

任务 16　使用默认及浮动路由实现网络连通

任务目标

1. 理解默认路由及浮动路由的作用。
2. 实现路由器和三层交换机的默认路由协议。
3. 实现路由器和三层交换机的浮动路由配置。

任务描述

某公司随着规模的不断扩大，现有北京总部和天津分部两个办公地点，天津分部与北京总部之间使用路由器连接。该公司的网络管理员小赵经过考虑，决定在北京总部和天津分部之间的路由器上配置默认路由和浮动路由，减少网络管理，提高链路的可用性，使所有计算机能够互相访问。

配置浮动路由能够实现在北京总部与天津分部的主链路断开时，通过备份链路连接。

任务要求

（1）使用默认路由及浮动路由实现网络连通，网络拓扑结构如图 16.1 所示。

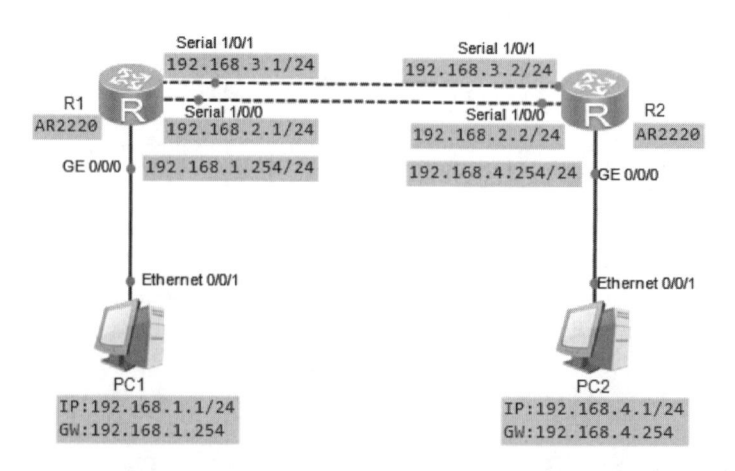

图 16.1　使用默认路由及浮动路由实现网络连通的网络拓扑结构

（2）路由器的端口及 IP 地址设置如表 16.1 所示。

表 16.1 路由器的端口及 IP 地址设置

设 备 名 称	端　　　口	IP 地址/子网掩码
R1	Serial 1/0/0	192.168.2.1/24
	Serial 1/0/1	192.168.3.1/24
	GE 0/0/0	192.168.1.254/24
R2	Serial 1/0/0	192.168.2.2/24
	Serial 1/0/1	192.168.3.2/24
	GE 0/0/0	192.168.4.254/24

（3）PC1、PC2 的端口及 IP 地址设置如表 16.2 所示。

表 16.2 PC1、PC2 的端口及 IP 地址设置

设 备 名 称	本端端口	IP 地址/子网掩码	网　　关	对端设备与端口
PC1	Ethernet 0/0/1	192.168.1.1/24	192.168.1.254	R1:GE 0/0/0
PC2	Ethernet 0/0/1	192.168.4.1/24	192.168.4.254	R2:GE 0/0/0

（4）通过在两台路由器上添加默认路由及浮动路由，实现全网互通和链路备份。在配置浮动路由优先级时，配置 192.168.2.0 网段为主链路，192.168.3.0 网段为备份链路，最终实现北京总部计算机与天津分部计算机的连通。

任务实施

步骤 1：按照图 16.1 搭建网络拓扑，在路由器的 Serial 1/0/0 端口位置上添加 2SA 模块，路由器之间的连线使用 Serial 串口线，其他使用直通线，开启所有设备电源，并为每台计算机设置相应的 IP 地址和子网掩码。

步骤 2：对路由器进行基本配置。

（1）对路由器 R1 进行基本配置。

```
<Huawei>system-view
[Huawei]sysname R1
[R1]interface GigabitEthernet 0/0/0
[R1-GigabitEthernet0/0/0]ip add 192.168.1.254 24
[R1-GigabitEthernet0/0/0]quit
[R1]interface Serial 1/0/0
[R1-Serial1/0/0]ip add 192.168.2.1 24
[R1-Serial1/0/0]quit
[R1]interface Serial 1/0/1
[R1-Serial1/0/1]ip add 192.168.3.1 24
[R1-Serial1/0/1]quit
```

（2）对路由器 R2 进行基本配置。

```
<Huawei>system-view
[Huawei]sysname R2
[R2]interface GigabitEthernet 0/0/0
[R2-GigabitEthernet0/0/0]ip add 192.168.4.254 24
[R2-GigabitEthernet0/0/0]quit
[R2]interface Serial 1/0/0
[R2-Serial1/0/0]ip add 192.168.2.2 24
[R2-Serial1/0/0]quit
[R2]interface Serial 1/0/1
[R2-Serial1/0/1]ip add 192.168.3.2 24
[R2-Serial1/0/1]quit
```

步骤 3：配置默认路由，实现全网互通。

（1）在路由器 R1 上进行配置。

```
[R1]ip route-static 0.0.0.0 0.0.0.0 192.168.2.2         //配置默认路由
```

（2）在路由器 R2 上进行配置。

```
[R2]ip route-static 0.0.0.0 0.0.0.0 192.168.2.1         //配置默认路由
```

步骤 4：配置浮动静态路由，实现链路备份。

（1）在路由器 R1 上进行配置。

```
[R1]ip route-static 0.0.0.0 0.0.0.0 192.168.3.2 preference 90    //修改优先级为 90
```

（2）在路由器 R2 上进行配置。

```
[R2]ip route-static 0.0.0.0 0.0.0.0 192.168.3.1 preference 90    //修改优先级为 90
```

任务验收

（1）在路由器 R1 上，使用"display ip routing-table"命令查看路由表。

```
[R1]display ip routing-table
Route Flags: R - relay, D - download to fib
------------------------------------------------------------------------------
Routing Tables: Public
        Destinations : 16       Routes : 16

Destination/Mask    Proto   Pre  Cost   Flags NextHop         Interface
      0.0.0.0/0     Static  60   0      RD    192.168.2.2     Serial1/0/0
      127.0.0.0/8   Direct  0    0      D     127.0.0.1       InLoopBack0
      127.0.0.1/32  Direct  0    0      D     127.0.0.1       InLoopBack0
127.255.255.255/32  Direct  0    0      D     127.0.0.1       InLoopBack0
    192.168.1.0/24  Direct  0    0      D     192.168.1.254   GigabitEthernet0/0/0
  192.168.1.254/32  Direct  0    0      D     127.0.0.1       GigabitEthernet0/0/0
```

192.168.1.255/32	Direct	0	0	D	127.0.0.1	GigabitEthernet0/0/0
192.168.2.0/24	Direct	0	0	D	192.168.2.1	Serial1/0/0
192.168.2.1/32	Direct	0	0	D	127.0.0.1	Serial1/0/0
192.168.2.2/32	Direct	0	0	D	192.168.2.2	Serial1/0/0
192.168.2.255/32	Direct	0	0	D	127.0.0.1	Serial1/0/0
192.168.3.0/24	Direct	0	0	D	192.168.3.1	Serial1/0/1
192.168.3.1/32	Direct	0	0	D	127.0.0.1	Serial1/0/1
192.168.3.2/32	Direct	0	0	D	192.168.3.2	Serial1/0/1
192.168.3.255/32	Direct	0	0	D	127.0.0.1	Serial1/0/1
255.255.255.255/32	Direct	0	0	D	127.0.0.1	InLoopBack0

（2）使用 PC1 ping PC2 的 IP 地址测试连通性，连通性测试结果如图 16.2 所示，可以看到是连通的。

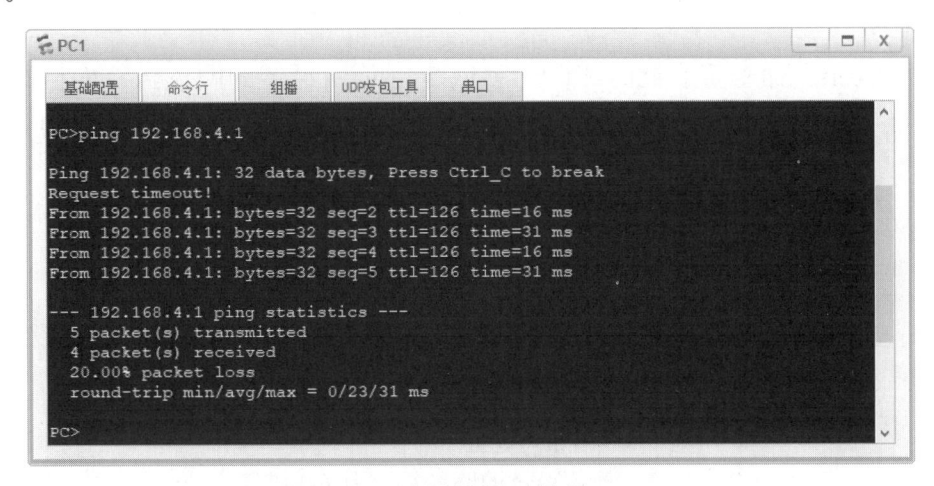

图 16.2　连通性测试结果（1）

（3）使用"tracert"命令查看此时 PC1 与 PC2 通信所经过的网关，从而检测经过的路径是否为主链路，如图 16.3 所示。

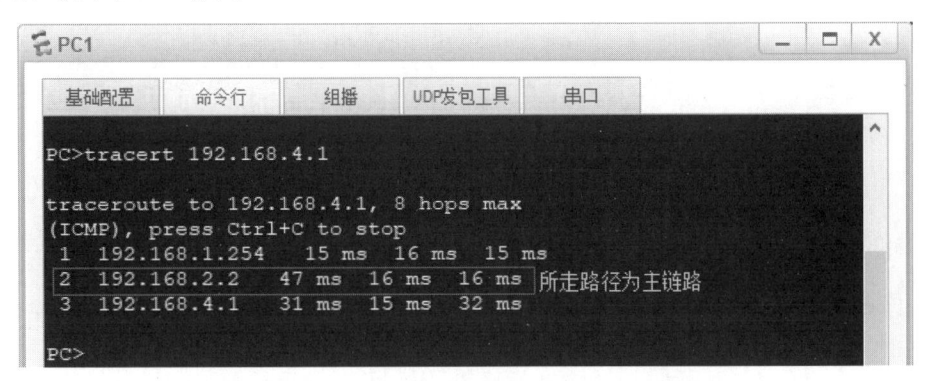

图 16.3　检测经过的路径是否为主链路

（4）测试计算机通信时使用的备用链路。

① 将路由器 R1 的 Serial 1/0/0 端口关闭。使用 PC1 ping PC2 的 IP 地址测试连通性，连通性测试结果如图 16.4 所示，可以看到短暂的超时后，依然是连通的。

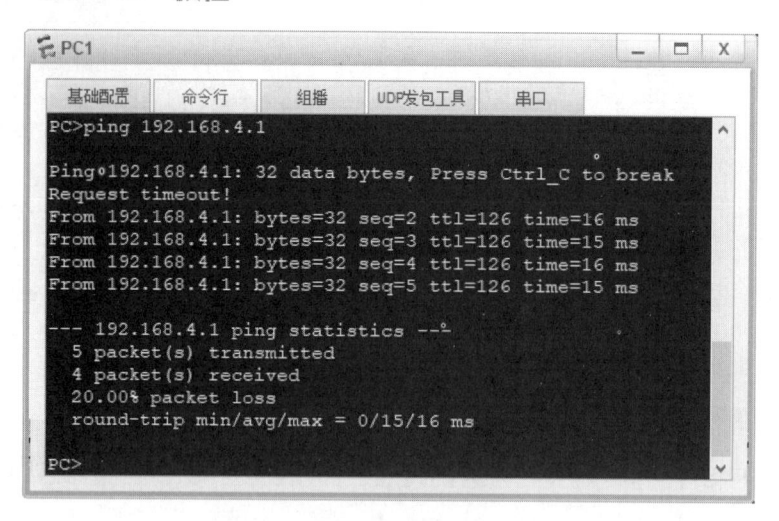

图16.4 连通性测试结果（2）

② 使用"tracert"命令查看此时 PC1 与 PC2 通信所经过的网关，从而检测经过的路径是否为备用链路，如图 16.5 所示。

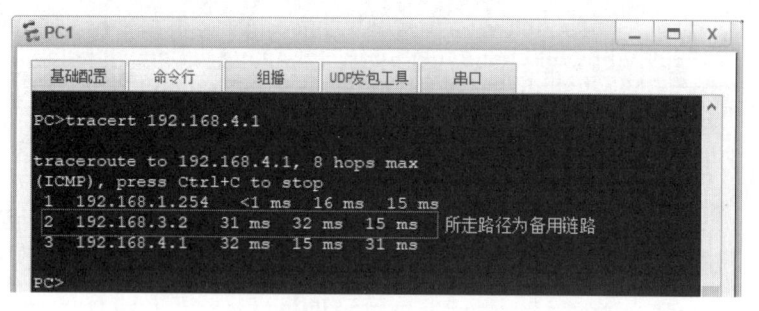

图16.5 检测经过的路径是否为备用链路

任务 17　使用动态路由 RIPv2 协议实现网络连通

任务目标

1．理解静态路由和动态路由的区别。
2．实现动态路由 RIPv2 协议和 MD5 认证的配置。

任务描述

某公司随着规模的不断扩大，路由器的数量开始增加。网络管理员小赵发现原有的静态路由已经不适合现在的公司，需要配置动态路由 RIPv2 协议，从而实现网络中所有主机之间的连通。

在路由器较多的网络环境中，手动配置静态路由会给网络管理员带来很大的工作负担，使用动态路由 RIPv2 协议可以很好地解决此问题。同时为了提高安全性，需要在路由器和交换机之间配置 MD5 认证。

任务要求

（1）使用动态路由 RIPv2 协议实现网络连通，网络拓扑结构如图 17.1 所示。

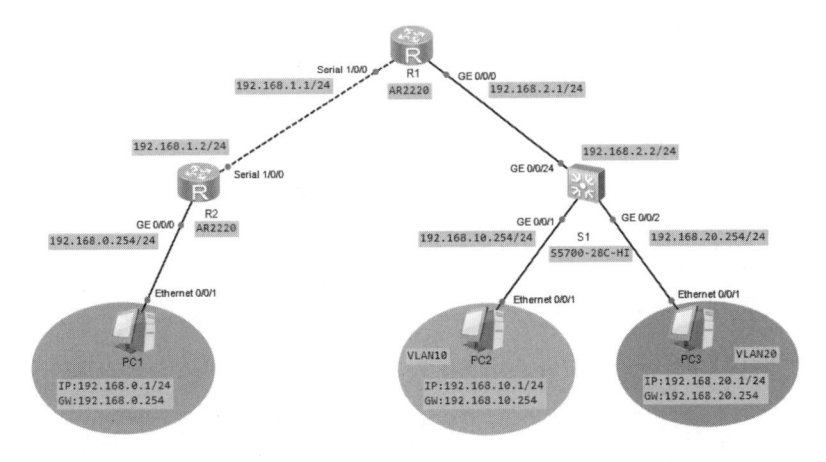

图 17.1　使用动态路由器 RIPv2 协议实现网络连通的网络拓扑结构

（2）路由器和交换机的端口及 IP 地址设置如表 17.1 所示。

表 17.1　路由器和交换机的端口及 IP 地址设置

设备名称	端口	IP 地址/子网掩码
R1	GE 0/0/0	192.168.2.1/24
	Serial 1/0/0	192.168.1.1/24
R2	Serial 1/0/0	192.168.1.2/24
	GE 0/0/0	192.168.0.254/24
S1	GE 0/0/1（VLANIF10）	192.168.10.254/24
	GE 0/0/2（VLANIF20）	192.168.20.254/24
	GE 0/0/24（VLANIF30）	192.168.2.2/24

（3）PC1～PC3 的端口及 IP 地址设置如表 17.2 所示。

表 17.2　PC1～PC3 的端口及 IP 地址设置

设备名称	本端端口	IP 地址/子网掩码	网关	对端设备与端口
PC1	Ethernet 0/0/1	192.168.0.1/24	192.168.0.254	R2:GE 0/0/0
PC2	Ethernet 0/0/1	192.168.10.1/24	192.168.10.254	S1:GE 0/0/1
PC3	Ethernet 0/0/1	192.168.20.1/24	192.168.20.254	S1:GE 0/0/2

（4）配置动态路由 RIPv2 协议和 MD5 的认证，密码为"huawei"，实现网络连通。

任务实施

步骤 1：按照图 17.1 搭建网络拓扑，在路由器的 Serial 1/0/0 端口位置上添加 2SA 模块，路由器之间的连线使用 Serial 串口线，其他使用直通线，开启所有设备电源，并为每台计算机设置相应的 IP 地址和子网掩码。

步骤 2：设置交换机和路由器的端口及 IP 地址等参数。

设置交换机和路由器的端口及 IP 地址参数，具体请参照任务 15 中的交换机 S1、路由器 R1 和路由器 R2 的基本配置。

步骤 3：配置 RIPv2 协议，实现网络连通。

（1）对交换机 S1 进行路由配置。

交换机 S1 上直连的网络有 192.168.1.0、192.168.10.0 和 192.168.20.0，因此要添加的 RIP 路由协议如下。

```
[S1]rip                              //进入 RIP 路由协议
[S1-rip-1]version 2                  //配置为 RIPv2 协议
[S1-rip-1]network 192.168.2.0        //通告直连网络
[S1-rip-1]network 192.168.10.0       //通告直连网络
[S1-rip-1]network 192.168.20.0       //通告直连网络
```

```
[S1-rip-1]quit
```

（2）对路由器 R1 进行路由配置。

与路由器 R1 直连的网络有 192.168.1.0 和 192.168.2.0，因此要添加的 RIP 路由如下。

```
[R1]rip                               //进入 RIP 路由协议
[R1-rip-1]version 2                   //配置为 RIPv2 协议
[R1-rip-1]network 192.168.1.0         //通告直连网络
[R1-rip-1]network 192.168.2.0         //通告直连网络
[R1-rip-1]quit
```

（3）对路由器 R2 进行路由配置。

与路由器 R2 直连的网络有 192.168.0.0 和 192.168.1.0，因此要添加的 RIP 路由如下。

```
[R2]rip                               //进入 RIP 路由协议
[R2-rip-1]version 2                   //配置为 RIPv2 协议
[R2-rip-1]network 192.168.0.0         //通告直连网络
[R2-rip-1]network 192.168.1.0         //通告直连网络
[R2-rip-1]quit
```

步骤 4：配置 MD5 认证。

（1）配置交换机 S1。

```
[S1]int vlan 30
[S1-Vlanif30]rip authentication-mode md5 usual huawei//使用 MD5 通用报文格式密文认证
```

（2）配置路由器 R1。

```
[R1]int serial 1/0/0
[R1-Serial1/0/0]rip authentication-mode md5 usual huawei
[R1-Serial1/0/0]int GigabitEthernet 0/0/0
[R1-GigabitEthernet0/0/0]rip authentication-mode md5 usual huawei
```

（3）配置路由器 R2。

```
[R2]int serial 1/0/0
[R2-Serial1/0/0]rip authentication-mode md5 usual huawei
```

任务验收

（1）在路由器 R1 上，使用"display ip routing-table"命令查看路由表。

```
[R1]display ip routing-table
Route Flags: R - relay, D - download to fib
------------------------------------------------------------------
Routing Tables: Public
       Destinations : 14      Routes : 14
Destination/Mask        Proto   Pre  Cost    Flags NextHop      Interface
127.0.0.0/8             Direct  0    0         D   127.0.0.1    InLoopBack0
127.0.0.1/32            Direct  0    0         D   127.0.0.1    InLoopBack0
```

```
    127.255.255.255/32       Direct   0    0    D   127.0.0.1     InLoopBack0
    192.168.0.0/24           RIP      100  1    D   192.168.1.2   Serial1/0/0
    192.168.1.0/24           Direct   0    0    D   192.168.1.1   Serial1/0/0
    192.168.1.1/32           Direct   0    0    D   127.0.0.1     Serial1/0/0
    192.168.1.2/32           Direct   0    0    D   192.168.1.2   Serial1/0/0
    192.168.1.255/32         Direct   0    0    D   127.0.0.1     Serial1/0/0
    192.168.2.0/24           Direct   0    0    D   192.168.2.1   GigabitEthernet0/0/0
    192.168.2.1/32           Direct   0    0    D   127.0.0.1     GigabitEthernet0/0/0
    192.168.2.255/32         Direct   0    0    D   127.0.0.1     GigabitEthernet0/0/0
    192.168.10.0/24          RIP      100  1    D   192.168.2.2   GigabitEthernet0/0/0
    192.168.20.0/24          RIP      100  1    D   192.168.2.2   GigabitEthernet0/0/0
    255.255.255.255/32       Direct   0    0    D   127.0.0.1     InLoopBack0
```

（2）在路由器 R1 上，使用"display rip 1 interface g0/0/0 verbose"命令查看 GE 0/0/0 端口的认证类型[①]。

```
<R1>display rip 1 interface g0/0/0 verbose
 GigabitEthernet0/0/0(192.168.2.1)
  State                : UP              MTU        : 500
  Metricin             : 0
  Metricout            : 1
  Input                : Enabled    Output     : Enabled
  Protocol             : RIPv2 Multicast
  Send version         : RIPv2 Multicast Packets
  Receive version      : RIPv2 Multicast and Broadcast Packets
  Poison-reverse       : Disabled
  Split-Horizon        : Enabled
  Authentication type  : MD5 (Usual)
  Replay Protection    : Disabled
```

（3）使用 PC1 ping PC2 和 PC3 的 IP 地址，通过连通性测试结果可以看到三者是连通的。

---

① 在本书中，g 表示 GigabitEthernet。

任务 18　使用动态路由 OSPF 实现网络连通

任务目标

1. 理解 OSPF 的工作原理。
2. 实现动态路由 OSPF 和 MD5 认证的配置。

任务描述

某公司随着规模的不断扩大，路由器的数量在原有的基础上有所增加。网络管理员小赵发现原有的路由协议已经不适合现有的网络环境，可以配置动态路由 OSPF（Open Shortest Path First，开放最短通路优先协议），从而实现网络中所有主机之间的互相通信。动态路由 OSPF 可以实现快速收敛，并且出现环路的可能性不大，适合中型和大型企业网络。同时为了提高安全性，在路由器和交换机之间配置 MD5 认证。

任务要求

（1）使用动态路由 OSPF 以实现网络连通，网络拓扑结构如图 18.1 所示。

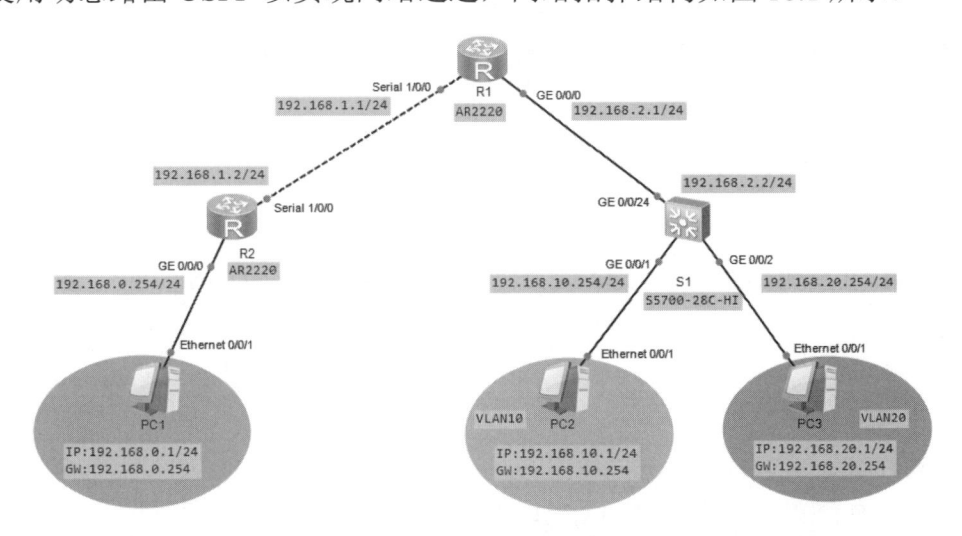

图 18.1　使用动态路由 OSPF 以实现网络连通的网络拓扑结构

（2）路由器和交换机的端口及 IP 地址设置如表 18.1 所示。

表 18.1 路由器和交换机的端口及 IP 地址设置

设备名称	端口	IP 地址/子网掩码
R1	GE 0/0/0	192.168.2.1/24
	Serial 1/0/0	192.168.1.1/24
R2	Serial 1/0/0	192.168.1.2/24
	GE 0/0/0	192.168.0.254/24
S1	GE 0/0/1（VLANIF10）	192.168.10.254/24
	GE 0/0/2（VLANIF20）	192.168.20.254/24
	GE 0/0/24（VLANIF30）	192.168.2.2/24

（3）PC1~PC3 的端口及 IP 地址设置如表 18.2 所示。

表 18.2 PC1~PC3 的端口及 IP 地址设置

设备名称	本端端口	IP 地址/子网掩码	网关	对端设备与端口
PC1	Ethernet0/0/1	192.168.0.1/24	192.168.0.254	R2:GE 0/0/0
PC2	Ethernet0/0/1	192.168.10.1/24	192.168.10.254	S1:GE 0/0/1
PC3	Ethernet0/0/1	192.168.20.1/24	192.168.20.254	S1:GE 0/0/2

（4）配置动态路由 OSPF 和 MD5 的认证，密码为"huawei"，实现网络连通。

任务实施

步骤 1：按照图 18.1 搭建网络拓扑，在路由器的 Serial 1/0/0 端口位置上添加 2SA 模块，路由器之间的连线使用 Serial 串口线，其他使用直通线，开启所有设备电源，并为每台计算机设置相应的 IP 地址和子网掩码。

步骤 2：配置交换机和路由器的端口 IP 地址等参数。

配置交换机和路由器的端口 IP 地址等参数，具体请参照任务 15 中的交换机 S1、路由器 R1 和路由器 R2 的基本配置。

步骤 3：配置动态路由 OSPF，实现网络连通。

（1）配置交换机 S1。

与交换机 S1 直连的网络有 192.168.2.0、192.168.10.0 和 192.168.20.0，因此要添加如下的动态路由 OSPF。

```
[S1]ospf 1                                          //进入OSPF，1表示进程号，默认为1
[S1-ospf-1]area 0                                   //指定骨干区域0
[S1-ospf-1-area-0.0.0.0]network 192.168.2.0 0.0.0.255   //通告直连网络
[S1-ospf-1-area-0.0.0.0]network 192.168.10.0 0.0.0.255  //通告直连网络
[S1-ospf-1-area-0.0.0.0]network 192.168.20.0 0.0.0.255  //通告直连网络
```

```
[S1-ospf-1-area-0.0.0.0]return
```

（2）配置路由器 R1。

与路由器 R1 直连的网络有 192.168.1.0 和 192.168.2.0，因此要添加如下的动态路由 OSPF。

```
[R1]ospf 1
[R1-ospf-1]area 0
[R1-ospf-1-area-0.0.0.0]network 192.168.1.0 0.0.0.255
[R1-ospf-1-area-0.0.0.0]network 192.168.2.0 0.0.0.255
[R1-ospf-1-area-0.0.0.0]return
```

（3）配置路由器R2。

与路由器 R2 直连的网络有 192.168.0.0 和 192.168.1.0，因此要添加如下的动态路由 OSPF。

```
[R2]ospf 1
[R2-ospf-1]area 0
[R2-ospf-1-area-0.0.0.0]network 192.168.0.0 0.0.0.255
[R2-ospf-1-area-0.0.0.0]network 192.168.1.0 0.0.0.255
[R2-ospf-1-area-0.0.0.0]return
```

步骤 4：配置 MD5 认证。

OSPF 支持两种报文认证方式，分别是区域认证及端口认证。

1. 区域认证

（1）配置交换机 S1。

```
[S1]ospf 1
[S1-ospf-1]area 0
[S1-ospf-1-area-0.0.0.0]authentication-mode md5 1 cipher huawei
//将设备上所有属于 Area0 的端口激活 OSPF 报文认证功能，而且使用 MD5 的认证方式
```

（2）配置路由器 R1。

```
[R1]ospf 1
[R1-ospf-1]area 0
[R1-ospf-1-area-0.0.0.0]authentication-mode md5 1 cipher huawei
```

（3）配置路由器 R2。

```
[R2]ospf 1
[R2-ospf-1]area 0
[R2-ospf-1-area-0.0.0.0]authentication-mode md5 1 cipher huawei
```

2. 端口认证

（1）配置交换机 S1。

```
[S1]int vlan 30
[S1-Vlanif30]ospf authentication-mode md5 1 cipher huawei//使用 MD5 通用报文格式密文认证
```

（2）配置路由器 R1。

```
[R1]int serial 1/0/0
[R1-Serial1/0/0]ospf authentication-mode md5 1 cipher huawei
[R1-Serial1/0/0]int GigabitEthernet 0/0/0
[R1-GigabitEthernet0/0/0]ospf authentication-mode md5 1 cipher huawei
```

（3）配置路由器 R2。

```
[R2]int serial 1/0/0
[R2-Serial1/0/0]ospf authentication-mode md5 1 cipher huawei
```

任务验收

（1）在路由器 R1 上，使用"display ip routing-table protocol ospf"命令查看动态路由 OSPF 信息。

```
[R1]display ip routing-table protocol ospf
Route Flags: R - relay, D - download to fib
------------------------------------------------------------------------------
Public routing table : OSPF
        Destinations : 3        Routes : 3
OSPF routing table status : <Active>
        Destinations : 3        Routes : 3
Destination/Mask    Proto   Pre  Cost      Flags NextHop          Interface
   192.168.0.0/24   OSPF    10   49          D   192.168.1.2      Serial1/0/0
   192.168.10.0/24  OSPF    10   2           D   192.168.2.2      GigabitEthernet0/0/0
   192.168.20.0/24  OSPF    10   2           D   192.168.2.2      GigabitEthernet0/0/0
OSPF routing table status : <Inactive>
        Destinations : 0        Routes : 0
```

（2）使用 PC1 ping PC2 和 PC3 的 IP 地址，通过连通性测试结果可以看到三者是连通的。

任务 19　组建直连式二层无线局域网

任务目标

1. 正确完成 WLAN 的基础配置。
2. 实现 AC+AP 直连式二层组网配置。

任务描述

某公司构建了互通的办公网，现需要在网络中部署 WLAN 以满足员工的移动办公需求。考虑到消费级的无线路由器在性能、扩展性、管理性上都无法满足要求，网络管理员小赵准备采用 AC+AP 的方案。同时，为了不大幅度地增加部署的难度，小赵选择组建直连式二层无线局域网。

任务要求

（1）组建直连式二层无线局域网，网络拓扑结构如图 19.1 所示。

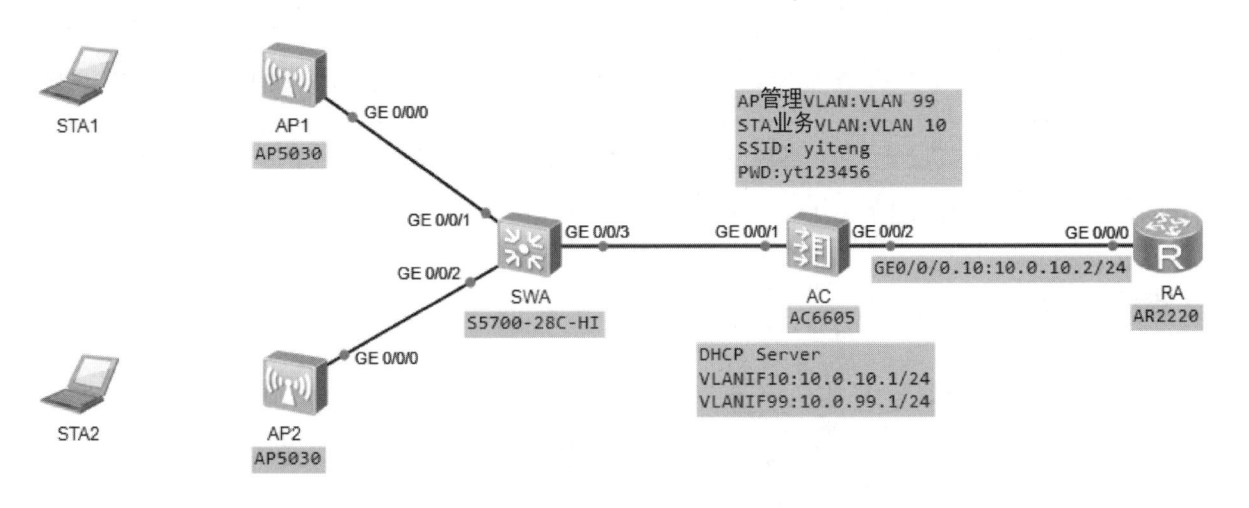

图 19.1　组建直连式二层无线局域网的网络拓扑结构

(2) 无线控制器 AC 数据规划如表 19.1 所示。

表 19.1 无线控制器 AC 数据规划

配 置 项	数 据
AP 管理 VLAN	VLAN99
STA 业务 VLAN	VLAN10
DHCP 服务器	无线控制器 AC 作为 DHCP 服务器为 AP 和 STA 分配 IP 地址
AP 的 IP 地址池	10.0.99.2～10.0.99.254/24
STA 的 IP 地址池	10.0.10.3～10.0.10.254/24
AC 的源端口 IP 地址	VLANIF99：10.0.99.1/24
AP 组	名称：ap-group1，引用模板：VAP 模板 wlan-net、域管理模板 default
域管理模板	名称：default，国家码：中国
SSID 模板	名称：wlan-net，SSID 名称：yiteng
安全模板	名称：wlan-net，安全策略：WPA-WPA2+PSK+AES，密码：yt123456
VAP 模板	名称：wlan-net，转发模式：直接转发，业务 VLAN：VLAN10，引用模板：SSID 模板 wlan-net、安全模板 wlan-net

(3) 路由器、交换机和无线控制器 AC 等网络设备端口及 IP 地址设置如表 19.2 所示。

表 19.2 路由器、交换机和无线控制器 AC 等网络设备端口及 IP 地址设置

设备名称	端 口	IP 地址/子网掩码	备 注
RA	GE 0/0/0.10	10.0.10.2/24	—
	LoopBack 0	10.10.10.10/24	—
AC	GE 0/0/2（VLANIF10）	10.0.10.1/24	业务 VLAN
	GE 0/0/1（VLANIF99）	10.0.99.1/24	—
SWA	GE 0/0/1（VLANIF99）	—	管理 VLAN
	GE 0/0/2（VLANIF99）	—	
AP1	GE 0/0/0	自动获取	—
AP2	GE 0/0/0	自动获取	—

(4) 组建直连式二层无线局域网，配置 AP 上线、WLAN 业务，在启动并连接 STA 后，各网络设备可以正确获取 IP 地址，各网络设备之间可以相互通信。

任务实施

步骤 1：对网络设备进行基础配置。

(1) 按照图 19.1 搭建网络拓扑，连线全部使用直通线，开启所有设备电源。

(2) 对交换机 SWA 进行基本配置。

```
<Huawei>system-view
```

```
[Huawei]sysname SWA
[SWA]vlan batch 10 99
[SWA]interface GigabitEthernet 0/0/1
[SWA-GigabitEthernet0/0/1]port link-type trunk
[SWA-GigabitEthernet0/0/1]port trunk pvid vlan 99            //剥离VLAN99数据标签转发
[SWA-GigabitEthernet0/0/1]port trunk allow-pass vlan 10 99   //允许VLAN10和VLAN99通过
[SWA-GigabitEthernet0/0/1]quit
[SWA]interface GigabitEthernet 0/0/2
[SWA-GigabitEthernet0/0/2]port link-type trunk
[SWA-GigabitEthernet0/0/2]port trunk pvid vlan 99            //剥离VLAN99数据标签转发
[SWA-GigabitEthernet0/0/2]port trunk allow-pass vlan 10 99   //允许VLAN10和VLAN99通过
[SWA-GigabitEthernet0/0/2]quit
[SWA]interface GigabitEthernet 0/0/3
[SWA-GigabitEthernet0/0/3]port link-type trunk
[SWA-GigabitEthernet0/0/3]port trunk allow-pass vlan 10 99
```

（3）对路由器 RA 进行基本配置。

```
<Huawei>system-view
[Huawei]sysname RA
[RA]interface GigabitEthernet 0/0/0.10
[RA-GigabitEthernet0/0/0.10]dot1q termination vid 10
[RA-GigabitEthernet0/0/0.10]ip address 10.0.10.2 24          //VLAN10的IP地址
[RA-GigabitEthernet0/0/0.10]arp broadcast enable
[RA-GigabitEthernet0/0/0.10]quit
[RA]interface LoopBack 0                                     //环回端口用于测试
[RA-LoopBack0]ip address 10.10.10.10 24                      //该地址模拟DNS服务器地址
[RA-LoopBack0]quit
[RA]ip route-static 10.0.99.0 24 10.0.10.1                   //通往VLAN99的静态路由
```

（4）对无线控制器 AC 进行基本配置。

```
<AC6605>system-view
[AC6605]sysname AC
[AC]vlan batch 10 99
[AC]interface GigabitEthernet 0/0/1
[AC-GigabitEthernet0/0/1]port link-type trunk
[AC-GigabitEthernet0/0/1]port trunk allow-pass vlan 10 99    //允许VLAN10和VLAN99通过
[AC-GigabitEthernet0/0/1]quit
[AC]interface GigabitEthernet 0/0/2
[AC-GigabitEthernet0/0/2]port link-type trunk
[AC-GigabitEthernet0/0/2]port trunk allow-pass vlan 10       //允许VLAN10通过
[AC-GigabitEthernet0/0/2]quit
[AC]interface Vlanif 10
[AC-Vlanif10]ip address 10.0.10.1 24                         //VLANIF10的端口地址
[AC-Vlanif10]interface Vlanif 99
[AC-Vlanif99]ip address 10.0.99.1 24                         //VLANIF99的端口地址
```

```
[AC-Vlanif99]quit
```

步骤 2：配置无线控制器 AC 上的 DHCP 服务器。

在无线控制器 AC 上配置 DHCP 服务器，为 STA 和 AP 动态分配 IP 地址。

```
[AC]dhcp enable                                              //开启 DHCP 服务
[AC]interface Vlanif 10
[AC-Vlanif10]dhcp select interface
[AC-Vlanif10]dhcp server excluded-ip-address 10.0.10.2
[AC-Vlanif10]dhcp server dns-list 10.10.10.10
[AC-Vlanif10]quit
[AC]interface Vlanif 99
[AC-Vlanif99]dhcp select interface
[AC-Vlanif99]quit
```

步骤 3：配置无线控制器 AC 上的默认路由。

```
[AC]ip route-static 0.0.0.0 0.0.0.0 10.0.10.2
```

步骤 4：查询 AP1 和 AP2 的 MAC 地址。

```
<AP1>display system-information
System Information
================================================
Serial Number        : 2102354483105F302471
System Time          : 2021-07-30 17:58:38
System Up time       : 55sec
System Name          : Huawei
Country Code         : US
MAC Address          : 00:e0:fc:76:25:F0
Radio 0 MAC Address  : 00:00:00:00:00:00
……                                                           //此处省略部分内容
//这里显示 AP1 的 MAC 地址为 00:e0:fc:76:25:F0
<AP2>display system-information
System Information
================================================
Serial Number        : 210235448310AE3FED5E
System Time          : 2021-07-30 18:00:14
System Up time       : 2min 24sec
System Name          : Huawei
Country Code         : US
MAC Address          : 00:e0:fc:c9:28:50
Radio 0 MAC Address  : 00:00:00:00:00:00
……                                                           //此处省略部分内容
//这里显示 AP2 的 MAC 地址为 00:e0:fc:c9:28:50
```

步骤 5：配置 AP 上线。

（1）创建 AP 组，用于将相同配置的 AP 加入同一 AP 组。

```
[AC]wlan                                            //进入 WLAN 视图
[AC-wlan-view]ap-group name ap-group1               //创建名为"ap-group1"的 AP 组
[AC-wlan-ap-group-ap-group1]quit
```

（2）创建域管理模板。在域管理模板中配置无线控制器 AC 的国家码，并引用域管理模板。

```
[AC-wlan-view]regulatory-domain-profile name default//创建并进入名为"default"的域管理模板
[AC-wlan-regulate-domain-default]country-code cn    //在域管理模板中配置无线控制器 AC 的国家
码为 cn
[AC-wlan-regulate-domain-default]quit
[AC-wlan-view]ap-group name ap-group1               //进入 ap-group1 AP 组
[AC-wlan-ap-group-ap-group1]regulatory-domain-profile default
                                                    //在 AP 组下引用刚建的 default 域管理模板
Warning: Modifying the country code will clear channel, power and antenna gain
configurations of the radio and reset the AP. Continue?[Y/N]:y
[AC-wlan-ap-group-ap-group1]quit
[AC-wlan-view]quit
```

（3）配置无线控制器 AC 的源端口。

```
[AC]capwap source interface Vlanif 99    //配置 AC 的源端口
```

（4）部署 AP，并配置无线控制器 AC 对 AP 的认证模式。

在无线控制器 AC 上，离线导入 AP1、AP2，AP 的 ID 分别为 0 和 1，并将 AP 加入 AP 监控组 ap-group1 中，部署 AP1、AP2 的名称分别为"office_1""office_2"，方便用户根据名称来查找 AP 的部署位置；配置无线控制器 AC 对 AP 的认证模式为 MAC 认证。

```
[AC]wlan
[AC-wlan-view]ap auth-mode mac-auth                 //配置无线控制器 AC 对 AP 的认证模式为 MAC 认证
[AC-wlan-view]ap-id 0 ap-mac 00e0-fc76-25f0         //通过 MAC 地址配置 AP1 的 ap-id 为 0
[AC-wlan-ap-0]ap-name office_1                      //部署 AP1 的名称为"office_1"
[AC-wlan-ap-0]ap-group ap-group1                    //将 AP1 加入 ap-group1 AP 监控组
Warning: This operation may cause AP reset. If the country code changes, it will
clear channel, power and antenna gain configurations of the radio, Whether to continue?
[Y/N]:y
[AC-wlan-ap-0]quit
[AC-wlan-view]ap-id 1 ap-mac 00e0-fcc9-2850
[AC-wlan-ap-1]ap-name office_2
[AC-wlan-ap-1]ap-group ap-group1
Warning: This operation may cause AP reset. If the country code changes, it will
clear channel, power and antenna gain configurations of the radio, Whether to continue?
[Y/N]:y
```

（5）在无线控制器 AC 上，使用"display ap all"命令，结果显示查看到 AP 的"State"字段为"nor"，表示 AP 正常上线。

```
<AC>display ap all                                  //查看 AP 上线状态
Info: This operation may take a few seconds. Please wait for a moment.done.
Total AP information:
```

```
nor : normal            [2]
--------------------------------------------------------------------
ID  MAC              Name     Group     IP         Type     State STA Uptime
--------------------------------------------------------------------
0   00e0-fc76-25f0   office_1  ap-group1  10.0.99.27 AP5030DN nor  0   27S
1   00e0-fcc9-2850   office_2  ap-group1  10.0.99.22 AP5030DN nor  0   39S
--------------------------------------------------------------------
Total:2                                              //显示总共2个AP
```

步骤6：配置WLAN业务。

（1）创建安全模板及配置安全策略，用于配置STA连接WLAN时使用的认证方式。

```
[AC]wlan
[AC-wlan-view]security-profile name wlan-net
                                    //创建名称为"wlan-net"的安全模板
[AC-wlan-sec-prof-wlan-net]security wpa-wpa2 psk pass-phrase yt123456 aes
                                    //配置安全策略
[AC-wlan-sec-prof-wlan-net]quit
```

（2）创建SSID模板，并配置SSID模板的名称。

```
[AC-wlan-view]ssid-profile name wlan-net       //创建名称为"wlan-net"的SSID模板
[AC-wlan-ssid-prof-wlan-net]ssid yiteng        //配置SSID模板的名称为"yiteng"
[AC-wlan-ssid-prof-wlan-net]quit
```

（3）创建VAP模板。创建名称为"wlan-net"的VAP模板，配置业务数据转发模式为直接转发，业务VLAN为VLAN 10，并且引用安全模板和SSID模板。

```
[AC-wlan-view]vap-profile name wlan-net        //创建名称为"wlan-net"的VAP模板
[AC-wlan-vap-prof-wlan-net]forward-mode direct-forward
                                    //配置业务数据转发模式为直接转发
[AC-wlan-vap-prof-wlan-net]service-vlan vlan-id 10
                                    //配置业务VLAN为VLAN10
[AC-wlan-vap-prof-wlan-net]security-profile wlan-net
                                    //引用安全模板wlan-net
[AC-wlan-vap-prof-wlan-net]ssid-profile wlan-net
                                    //引用SSID模板wlan-net
[AC-wlan-vap-prof-wlan-net]quit
```

（4）AP组引用VAP模板。配置AP的射频0和射频1都引用VAP模板wlan-net。

```
[AC-wlan-view]ap-group name ap-group1          //进入AP组ap-group1
[AC-wlan-ap-group-ap-group1]vap-profile wlan-net wlan 1 radio 0
                                    //射频0引用VAP模板
[AC-wlan-ap-group-ap-group1]vap-profile wlan-net wlan 1 radio 1
                                    //射频1引用VAP模板
[AC-wlan-ap-group-ap-group1]quit
```

（5）此时AP1和AP2上出现圆环状信号范围，表明AP1和AP2已经上线，如图19.2所示。

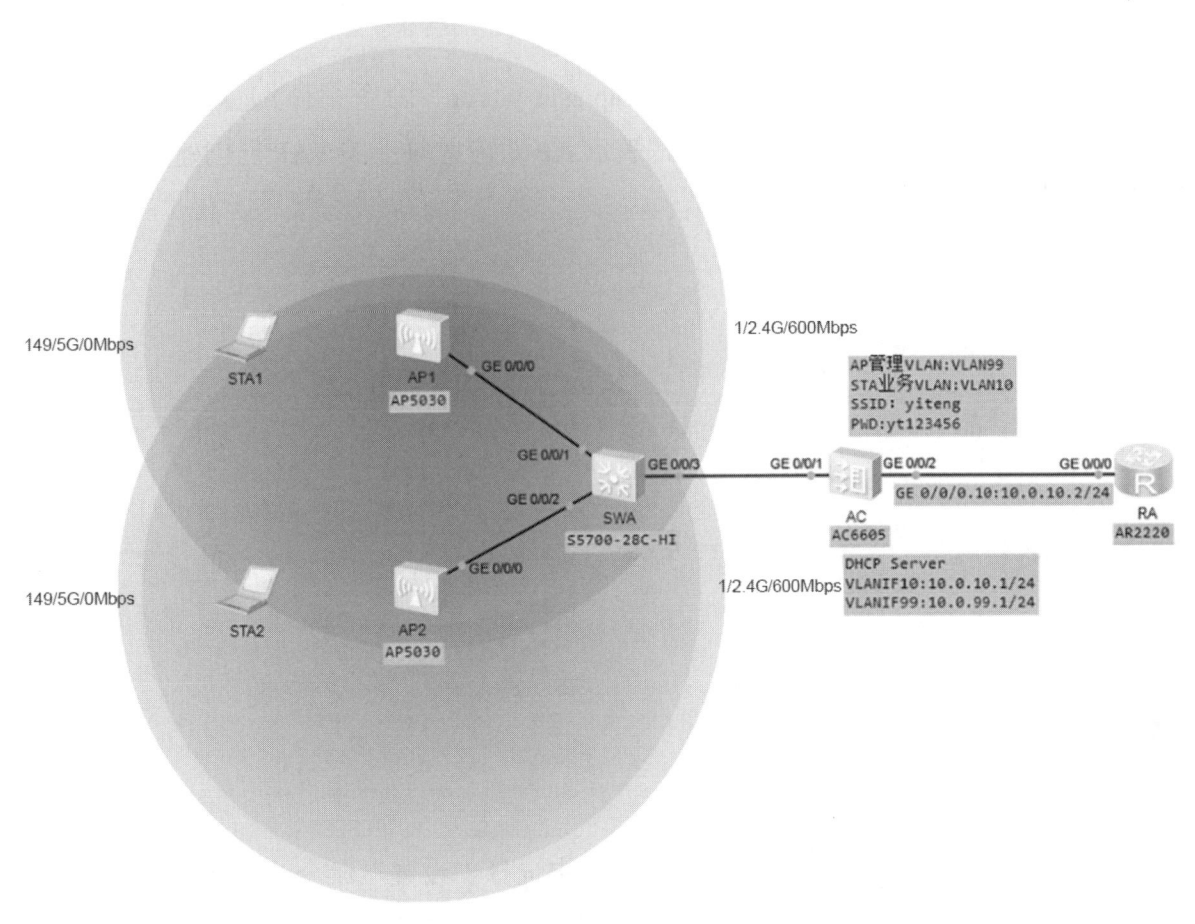

图 19.2　AP1 和 AP2 已经上线

（6）配置 AP 射频信道和功率。配置 AP1 射频 0 的带宽为 20MHz，信道为 6；配置 AP1 射频 1 的带宽为 20MHz，信道为 149。

```
[AC]wlan                                        //进入无线视图
[AC-wlan-view]ap-id 0                           //进入 AP1 视图
[AC-wlan-ap-0]radio 0                           //进入 AP1 射频 0 视图
[AC-wlan-radio-0/0]channel 20mhz 6              //配置射频 0 的带宽为 20MHz，信道为 6
Warning: This action may cause service interruption. Continue?[Y/N]y
[AC-wlan-ap-0]radio 1                           //进入 AP1 射频 1 视图
[AC-wlan-radio-0/1]channel 20mhz 149            //配置射频 1 的带宽为 20MHz，信道为 149
Warning: This action may cause service interruption. Continue?[Y/N]y
[AC-wlan-radio-0/1]quit
[AC-wlan-ap-0]quit
[AC-wlan-view]ap-id 1
[AC-wlan-ap-1]radio 0
[AC-wlan-radio-1/0]channel 20mhz 11
Warning: This action may cause service interruption. Continue?[Y/N]y
[AC-wlan-radio-1/0]quit
[AC-wlan-ap-1]radio 1
[AC-wlan-radio-1/1]channel 20mhz 153
Warning: This action may cause service interruption. Continue?[Y/N]y
```

```
[AC-wlan-ap-1]quit
```

（7）在无线控制器 AC 上使用"display vap ssid yiteng"命令查看 AP 对应射频上的 VAP 创建信息，当"Status"字段为"ON"时，表示 AP 对应射频上的 VAP 已创建成功。

```
<AC>display vap ssid yiteng                    //AP 射频上的 VAP 创建信息
Info: This operation may take a few seconds, please wait.
WID : WLAN ID
-----------------------------------------------------------------------------
AP ID  AP name   RfID WID BSSID         Status Auth type       STA  SSID
-----------------------------------------------------------------------------
0      office_1  0    1   00E0-FC76-25F0 ON    WPA/WPA2-PSK    0    yiteng
0      office_1  1    1   00E0-FC97-2600 ON    WPA/WPA2-PSK    0    yiteng
1      office_2  0    1   00E0-FCC9-2850 ON    WPA/WPA2-PSK    1    yiteng
1      office_2  1    1   00E0-FC7C-2860 ON    WPA/WPA2-PSK    0    yiteng
-----------------------------------------------------------------------------
Total: 4
```

任务验收

（1）无线客户端的测试。

启动 STA1，右击 STA1，在弹出的快捷菜单中选择"设置"命令，打开 STA1 设置对话框。在"Vap 列表"选项卡的"Vap 列表"选区中，选择信道为 6 的 VAP，如图 19.3 所示。

图 19.3　STA1 的"Vap 列表"选区

（2）单击"连接"按钮，在打开的对话框中输入密码"yt123456"，如图 19.4 所示，并单击"确定"按钮，这时可以看到状态显示为"已连接"，表示连接成功，如图 19.5 所示。

（3）查看 STA1 的 IP 地址。

在 STA1 正常关联无线网络 yiteng 后，使用"ipconfig"命令查看 STA1 通过无线网络自动获取的 IP 地址，如图 19.6 所示。

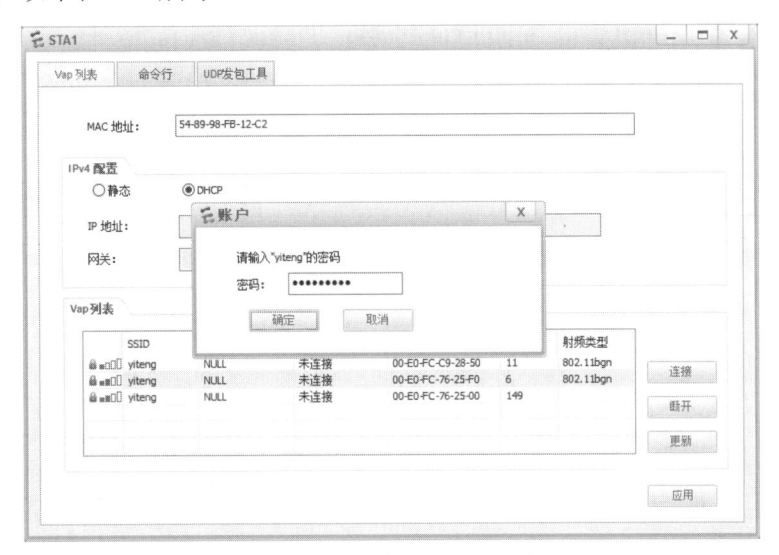

图 19.4　输入密码

图 19.5　状态显示为"已连接"

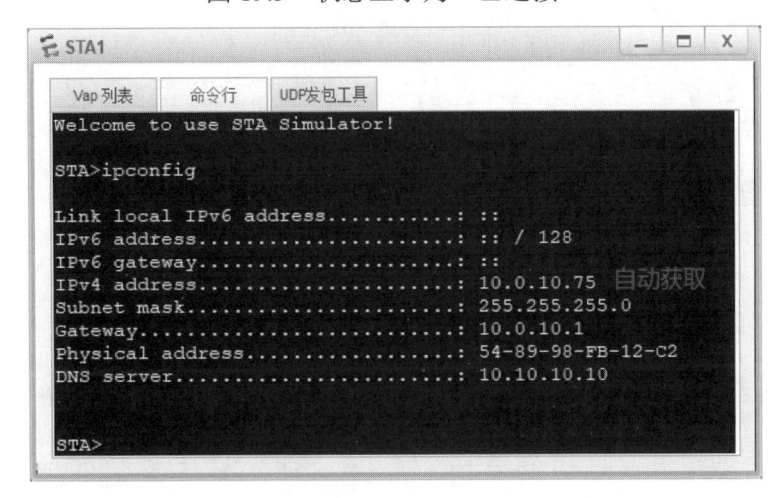

图 19.6　查看 STA1 自动获取的 IP 地址

（4）使用同样的方法连接 STA2。将设备全部连接无线网络后，在 AC 上执行"display station ssid yiteng"命令，可以查看到用户已经接入无线网络 yiteng 中。

```
<AC>display station ssid yiteng          //查看已接入无线网络 yiteng 中的用户
Rf/WLAN: Radio ID/WLAN ID
Rx/Tx: link receive rate/link transmit rate(Mbps)
--------------------------------------------------------------------------
STA MAC    AP ID Ap name  Rf/WLAN Band Type Rx/Tx  RSSI VLAN IP address
--------------------------------------------------------------------------
5489-9825-5712  1 office_2 0/1    2.4G  -    -/-    -   10   10.0.10.75
5489-98fb-12c2  1 office_2 0/1    2.4G  -    -/-    -   10   10.0.10.14
--------------------------------------------------------------------------
Total: 2 2.4G: 2 5G: 0                   //结果显示已接入的用户数为 2
```

（5）最终所有用户都可以获取正确的 IP 地址，实现全网互通，如图 19.7 所示。

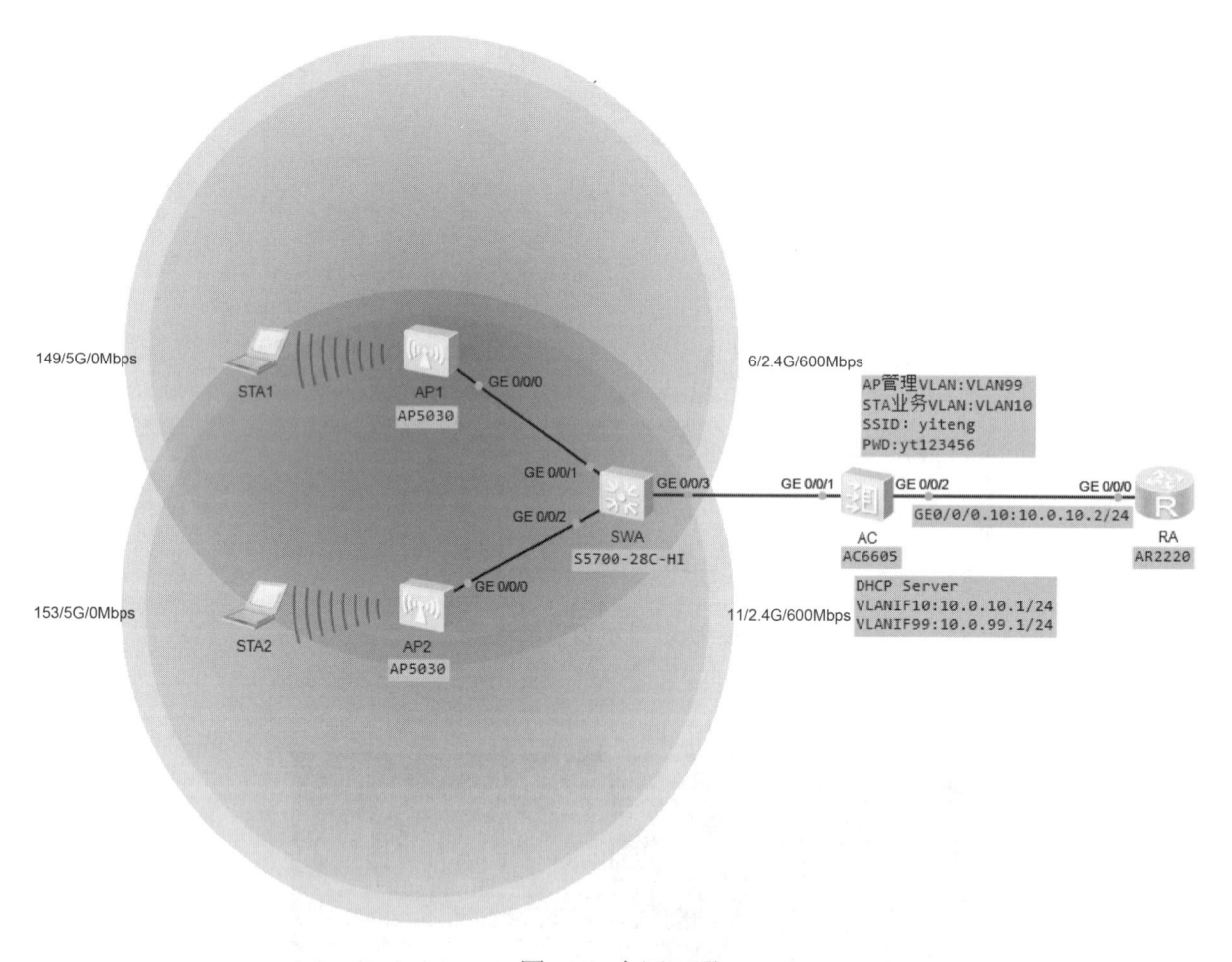

图 19.7 全网互通

（6）测试 STA1 与 DNS Server 及 VALN99 的网络连通性，在连通性测试结果中可以看到全网互通。

任务 20　组建旁挂式三层无线局域网

任务目标

1. 正确完成 WLAN 的安全配置。
2. 实现 AC+AP 旁挂式三层无线局域网配置。

任务描述

某公司需要在原有网络中部署 WLAN，以满足员工的移动办公需求。由于原来的有线网络较复杂，为满足 WLAN 组网的灵活性，网络管理员小赵准备采用"AC+瘦 AP 旁挂式三层无线局域网"方案，将 AP1 部署在销售部办公室，将 AP2 部署在财务部办公室。

任务要求

（1）组建旁挂式三层无线局域网，网络拓扑结构如图 20.1 所示。

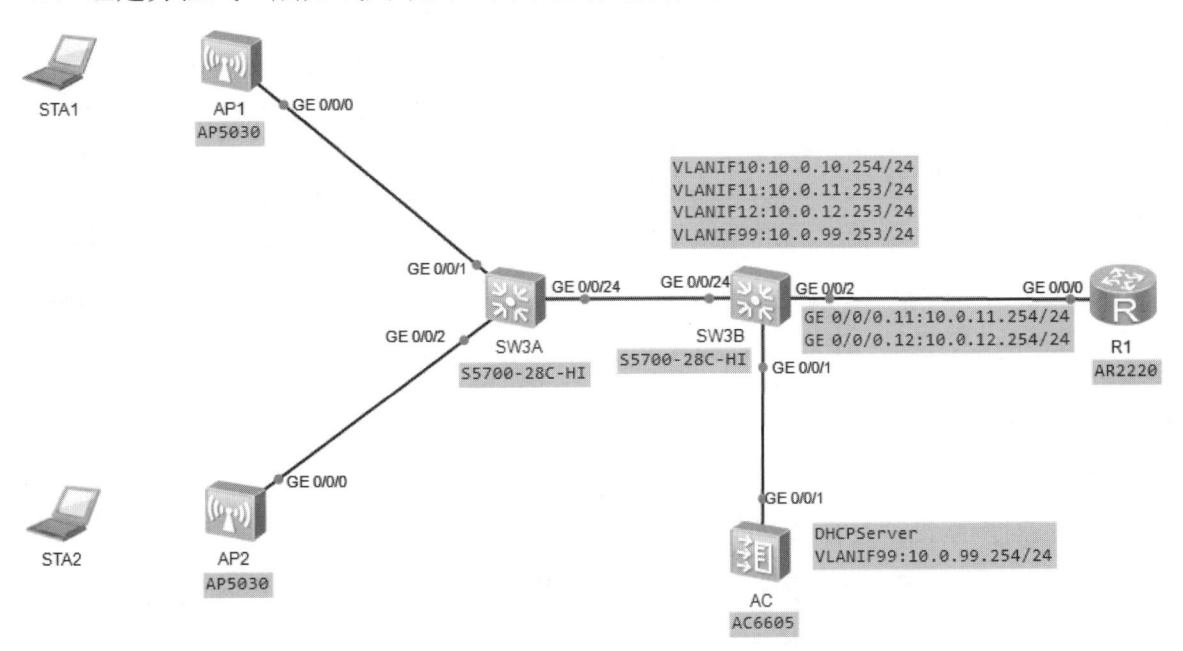

图 20.1　组建旁挂式三层无线局域网的网络拓扑结构

（2）无线控制器 AC 数据规划如表 20.1 所示。

表 20.1 无线控制器 AC 数据规划

配 置 项	数 据
DHCP 服务器	无线控制器 AC 作为 AP 和 STA 的 DHCP 服务器，可以汇聚交换机，以实现三层路由，STA 默认网关分别为 10.0.11.1、10.0.12.1
AP 的 IP 地址池	10.0.99.2~10.0.99.254/24
STA 的 IP 地址池	10.0.11.3~10.0.11.254/24、10.0.12.3~10.0.12.254/24
AC 的源端口	VLANIF99：10.0.99.2/24
AP 组	名称：ap-group1，引用模板：VAP 模板 wlan-net、域管理模板 default
域管理模板	名称：default，国家码：CN
SSID 模板	名称：wlan-net1、wlan-net2；SSID 名称：Sales、Finances
安全模板	名称：wlan-net1、wlan-net2；安全策略：WPA-WPA2+PSK+AES；密码：a1234567
VAP 模板	名称：wlan-net，转发模式：隧道转发，业务 VLAN：VLAN pool， 引用模板：SSID 模板 wlan-net、安全模板 wlan-net

（3）路由器、交换机和无线控制器 AC 等网络设备的端口及 IP 地址设置如表 20.2 所示。

表 20.2 路由器、交换机和无线控制器 AC 等网络设备的端口及 IP 地址设置

设备名称	端 口	IP 地址/子网掩码	备 注
R1	GE 0/0/0.11	10.0.11.254/24	—
	GE 0/0/0.12	10.0.12.254/24	—
AC	GE 0/0/1	10.0.99.254/24	—
SW3B	VLANIF10	10.0.10.254/24	管理 VLAN
	VLANIF99	10.0.99.253/24	
	VLANIF11	10.0.11.253/24	业务 VLAN
	VLANIF12	10.0.12.253/24	
AP1	GE 0/0/0	自动获取	Sales（销售部）
AP2	GE 0/0/0	自动获取	Finances（财务部）

（4）组建 AC+AP 旁挂式三层无线局域网：无线控制器 AC 作为 DHCP 服务器为 AP 和 STA 动态分配 IP 地址；交换机 SW3B 作为 DHCP 代理；采用隧道转发的业务数据转发方式。通过适当的配置实现 AP 上线、STA 正确获取 IP 地址，各网络设备之间可以相互通信。

任务实施

步骤 1：对网络设备进行基础配置。

（1）按照图 20.1 搭建网络拓扑，连线全部使用直通线，开启所有设备电源。

（2）对交换机 SW3A 进行基本配置。

```
[Huawei]sysname SW3A
[SW3A]vlan batch 10 to 12
[SW3A]interface GigabitEthernet 0/0/1
[SW3A-GigabitEthernet0/0/1]port link-type trunk
[SW3A-GigabitEthernet0/0/1]port trunk pvid vlan 10        //剥离VLAN10数据标签转发
[SW3A-GigabitEthernet0/0/1]port trunk allow-pass vlan 10 11 12
[SW3A-GigabitEthernet0/0/1]quit
[SW3A]interface GigabitEthernet 0/0/2
[SW3A-GigabitEthernet0/0/2]port link-type trunk
[SW3A-GigabitEthernet0/0/2]port trunk pvid vlan 10
[SW3A-GigabitEthernet0/0/2]port trunk allow-pass vlan 10 11 12
[SW3A-GigabitEthernet0/0/2]quit
[SW3A]interface GigabitEthernet 0/0/24
[SW3A-GigabitEthernet0/0/24]port link-type trunk
[SW3A-GigabitEthernet0/0/24]port trunk allow-pass vlan 10 11 12
```

（3）对交换机SW3B进行基本配置。

```
[Huawei]sysname SW3B
[SW3B]vlan batch 10 11 12 99
[SW3B]interface GigabitEthernet 0/0/24
[SW3B-Ethernet0/0/24]port link-type trunk
[SW3B-Ethernet0/0/24]port trunk allow-pass vlan 10 11 12
[SW3B]interface GigabitEthernet0/0/1
[SW3B-GigabitEthernet0/0/1]port link-type trunk
[SW3B-GigabitEthernet0/0/1]port trunk allow-pass vlan 11 12 99
[SW3B]interface GigabitEthernet0/0/2
[SW3B-GigabitEthernet0/0/2]port link-type trunk
[SW3B-GigabitEthernet0/0/2]port trunk allow-pass vlan 11 12
[SW3B-GigabitEthernet0/0/2]quit
[SW3B]interface Vlanif 10
[SW3B-Vlanif10]ip address 10.0.10.254 24
[SW3B-Vlanif10]interface Vlanif 11
[SW3B-Vlanif11]ip address 10.0.11.253 24
[SW3B-Vlanif11]interface Vlanif 12
[SW3B-Vlanif12]ip address 10.0.12.253 24
[SW3B-Vlanif112]interface Vlanif 99
[SW3B-Vlanif99]ip address 10.0.99.253 24
[SW3B-Vlanif99]quit
```

（4）对路由器R1进行基本配置。

```
[Huawei]sysname R1
[R1]interface GigabitEthernet0/0/0.11
[R1-GigabitEthernet0/0/0.11]dot1q termination vid 11
[R1-GigabitEthernet0/0/0.11]ip address 10.0.11.254 24
[R1-GigabitEthernet0/0/0.11]arp broadcast enable
```

```
[R1-GigabitEthernet0/0/0.11]quit
[R1]interface GigabitEthernet0/0/0.12
[R1-GigabitEthernet0/0/0.12]dot1q termination vid 12
[R1-GigabitEthernet0/0/0.12]ip address 10.0.12.254 24
[R1-GigabitEthernet0/0/0.12]arp broadcast enable
[R1-GigabitEthernet0/VLAN0/0.12]quit
```

（5）对无线控制器 AC 进行基本配置。

```
[AC6605]sysname AC
[AC]vlan batch 11 12 99
[AC]interface GigabitEthernet0/0/1
[AC-GigabitEthernet0/0/1]port link-type trunk
[AC-GigabitEthernet0/0/1]port trunk allow-pass vlan 11 12 99
[AC-GigabitEthernet0/0/1]quit
[AC]interface Vlanif 99
[AC-Vlanif99]ip address 10.0.99.254 24
[AC-Vlanif99]quit
```

步骤 2：配置默认路由。

（1）配置无线控制器 AC 到 AP 的路由。

```
[AC]ip route-static 0.0.0.0 0.0.0.0 10.0.99.253          //指向 SW3B 的 VLAN99
```

（2）配置路由器 R1 的默认路由。

```
[R1]ip route-static 0.0.0.0 0.0.0.0 10.0.11.253          //指向 10.0.11.253
```

步骤 3：配置 DHCP 服务。

（1）在无线控制器 AC 上，配置 DHCP 服务器。

```
[AC]dhcp enable
[AC]ip pool huawei                                       //为 AP 提供地址
[AC-ip-pool-huawei]network 10.0.10.0 mask 24
[AC-ip-pool-huawei]gateway-list 10.0.10.254
[AC-ip-pool-huawei]option 43 sub-option 3 ascii 10.0.99.254 //指明 AC 的 IP 地址
[AC-ip-pool-huawei]quit
[AC]ip pool vlan11                                       //为销售部提供地址
[AC-ip-pool-vlan11]gateway-list 10.0.11.254
[AC-ip-pool-vlan11]network 10.0.11.0 mask 24
[AC-ip-pool-vlan11]dns-list 10.10.10.10
[AC-ip-pool-vlan11]quit
[AC]ip pool vlan12                                       //为财务部提供地址
[AC-ip-pool-vlan12]gateway-list 10.0.12.254
[AC-ip-pool-vlan12]network 10.0.12.0 mask 24
[AC-ip-pool-vlan12]dns-list 10.10.10.10
[AC-ip-pool-vlan12]quit
[AC]interface vlanif 99
[AC-Vlanif99]dhcp select global
```

（2）在交换机 SW3B 上，开启 DHCP 服务、配置 DHCP 中继，为 AP 和 STA 动态分配 IP 地址。

```
[SW3B]dhcp enable
[SW3B]interface Vlanif 10
[SW3B-Vlanif10]dhcp select relay                      //为 AP 分配 IP 地址
[SW3B-Vlanif10]dhcp relay server-ip 10.0.99.254
[SW3B-Vlanif10]quit
[SW3B]interface Vlanif 11
[SW3B-Vlanif11]dhcp select relay                      //为 STA1 分配 IP 地址
[SW3B-Vlanif11]dhcp relay server-ip 10.0.99.254
[SW3B-Vlanif11]interface Vlanif 12
[SW3B-Vlanif12]dhcp select relay                      //为 STA2 分配 IP 地址
[SW3B-Vlanif12]dhcp relay server-ip 10.0.99.254
```

步骤 4：查询 AP1 和 AP2 的 MAC 地址。

```
<AP1>display system-information
System Information
=================================================
Serial Number         : 210235448310C677AE0C
System Time           : 2021-07-30 18:03:48
System Up time        : 1min 3sec
System Name           : Huawei
Country Code          : US
MAC Address           : 00:e0:fc:c5:07:e0
Radio 0 MAC Address   : 00:00:00:00:00:00
……                                                    //此处省略部分内容
//这里显示 AP1 的 MAC 地址为 00:e0:fc:c5:07:e0
<AP2>display system-information
System Information
=================================================
Serial Number         : 210235448310AF323518
System Time           : 2021-07-30 18:05:22
System Up time        : 2min 31sec
System Name           : Huawei
Country Code          : US
MAC Address           : 00:e0:fc:03:04:90
Radio 0 MAC Address   : 00:00:00:00:00:00
……                                                    //此处省略部分内容
//这里显示 AP2 的 MAC 地址为 00:e0:fc:03:04:90
```

步骤 5：配置 AP 上线。

（1）创建 AP 组，用于将相同配置的 AP 加入同一 AP 组。

```
[AC]wlan
```

```
[AC-wlan-view]ap-group name ap-group1
```

（2）创建域管理模板，配置无线控制器 AC 的国家码，并在 AP 组下引用域管理模板。

```
[AC-wlan-view]regulatory-domain-profile name default
[AC-wlan-regulate-domain-default]country-code cn
[AC-wlan-regulate-domain-default]ap-group name ap-group1
[AC-wlan-ap-group-ap-group1]regulatory-domain-profile default
Warning: Modifying the country code will clear channel, power and antenna gain
configurations of the radio and reset the AP. Continue?[Y/N]:y
```

（3）配置无线控制器 AC 的源端口。

```
[AC]capwap source interface Vlanif 99
```

（4）在无线控制器 AC 上，离线导入 AP1、AP2，并将 AP 加入 AP 监控组 ap-group1 中。

```
[AC]wlan
[AC-wlan-view]ap auth-mode mac-auth                            //认证模式为MAC认证
[AC-wlan-view]ap-id 0 ap-mac 00e0-fcc5-07e0
[AC-wlan-ap-0]ap-name area_1                                   //AP1 的名称为"area_1"
[AC-wlan-ap-0]ap-group ap-group1
Warning: This operation may cause AP reset. If the country code changes, it will
 clear channel, power and antenna gain configurations of the radio, Whether to
continue? [Y/N]:y
[AC-wlan-ap-0]quit
[AC-wlan-view]ap-id 1 ap-mac 00e0-fc03-0490
[AC-wlan-ap-1]ap-name area_2                                   //AP2 的名称为"area_2"
[AC-wlan-ap-1]ap-group ap-group1
Warning: This operation may cause AP reset. If the country code changes, it will
 clear channel, power and antenna gain configurations of the radio, Whether to
continue? [Y/N]:y
```

（5）在无线控制器 AC 上，使用"display ap all"命令查看 AP 上线状态，结果显示 AP 的"State"字段为"nor"，表示 AP 正常上线。

```
<AC>display ap all                                             //查看AP上线状态
Info: This operation may take a few seconds. Please wait for a moment.done.
Total AP information:
nor : normal         [2]
--------------------------------------------------------------------------------
ID  MAC            Name    Group     IP            Type       State STA Uptime
--------------------------------------------------------------------------------
0   00e0-fcc5-07e0 area_1  ap-group1 10.0.10.236   AP5030DN   nor   0   3H:56M:5S
1   00e0-fc03-0490 area_2  ap-group1 10.0.10.210   AP5030DN   nor   0   3H:56M:11S
--------------------------------------------------------------------------------
Total: 2
```

步骤 6：配置 WLAN 业务。

（1）创建名称为"wlan-net1"的安全模板、SSID 模板和 VAP 模板，配置安全策略、密码、SSID 模板的名称和转发模式，并对模板进行引用。

```
[AC]wlan
[AC-wlan-view]security-profile name wlan-net1           //创建安全模板
[AC-wlan-sec-prof-wlan-net1]security wpa-wpa2 psk pass-phrase a1234567 aes //安全策略
[AC-wlan-sec-prof-wlan-net1]quit
[AC-wlan-view]ssid-profile name wlan-net1               //创建 SSID 模板
[AC-wlan-ssid-prof-wlan-net1]ssid Sales                 //配置 SSID 模板的名称
[AC-wlan-ssid-prof-wlan-net1]quit
[AC-wlan-view]vap-profile name wlan-net1                //创建 VAP 模板
[AC-wlan-vap-prof-wlan-net1]forward-mode tunnel         //转发模式为隧道模式
[AC-wlan-vap-prof-wlan-net1]service-vlan vlan-id 11     //业务 VLAN 为 VLAN11
[AC-wlan-vap-prof-wlan-net1]security-profile wlan-net1  //引用安全模板
[AC-wlan-vap-prof-wlan-net1]ssid-profile wlan-net1      //引用 SSID 模板
```

（2）创建名称为"wlan-net2"的安全模板、SSID 模板和 VAP 模板，配置安全策略、密码、SSID 模板的名称和转发模式，并对模板进行引用。

```
[AC]wlan
[AC-wlan-view]security-profile name wlan-net2           //创建安全模板
[AC-wlan-sec-prof-wlan-net2]security wpa-wpa2 psk pass-phrase a1234567 aes
[AC-wlan-sec-prof-wlan-net2]quit
[AC-wlan-view]ssid-profile name wlan-net2               //创建 SSID 模板
[AC-wlan-ssid-prof-wlan-net2]ssid Finances              //配置 SSID 模板的名称
[AC-wlan-ssid-prof-wlan-net2]quit
[AC-wlan-view]vap-profile name wlan-net2                //创建 VAP 模板
[AC-wlan-vap-prof-wlan-net2]forward-mode tunnel         //转发模式为隧道模式
[AC-wlan-vap-prof-wlan-net2]service-vlan vlan-id 12     //业务 VLAN 为 VLAN12
[AC-wlan-vap-prof-wlan-net2]security-profile wlan-net2  //引用安全模板
[AC-wlan-vap-prof-wlan-net2]ssid-profile wlan-net2      //引用 SSID 模板
```

（3）配置 AP 组引用 VAP 模板，AP 上射频 0 和射频 1 同时使用 VAP 模板"wlan-net1"和"wlan-net2"的配置。

```
[AC]wlan
[AC-wlan-view]ap-group name ap-group1
[AC-wlan-ap-group-ap-group1]vap-profile wlan-net1 wlan 1 radio 0
[AC-wlan-ap-group-ap-group1]vap-profile wlan-net1 wlan 1 radio 1
[AC-wlan-ap-group-ap-group1]vap-profile wlan-net2 wlan 2 radio 0
[AC-wlan-ap-group-ap-group1]vap-profile wlan-net2 wlan 2 radio 1
```

（4）配置 AP1 射频的带宽和信道。

```
[AC-wlan-view]ap-id 0
[AC-wlan-ap-0]radio 0
[AC-wlan-radio-0/0]channel 20mhz 6                      //配置射频 0 的带宽为 20MHz，信道为 6
```

```
Warning: This action may cause service interruption. Continue?[Y/N]y
[AC-wlan-radio-0/0]quit
[AC-wlan-ap-0]radio 1
[AC-wlan-radio-0/1]channel 20mhz 149          //配置射频1的带宽为20MHz，信道为149
Warning: This action may cause service interruption. Continue?[Y/N]y
```

（5）配置 AP2 射频的带宽和信道。注意，当 AP1 和 AP2 的信号覆盖范围重叠时，信道值需要有一定的间隔。

```
[AC-wlan-view]ap-id 1
[AC-wlan-ap-1]radio 0
[AC-wlan-radio-1/0]channel 20mhz 11
Warning: This action may cause service interruption. Continue?[Y/N]y
[AC-wlan-radio-1/0]eirp 127
[AC-wlan-radio-1/0]quit
[AC-wlan-ap-1]radio 1
[AC-wlan-radio-1/1]channel 20mhz 153
Warning: This action may cause service interruption. Continue?[Y/N]y
[AC-wlan-radio-1/1]eirp 127
```

任务验收

（1）在无线控制器 AC 上使用 "display vap ssid Sales" 命令，查看 AP 对应射频上的 VAP 创建信息，当 "Status" 字段为 "ON" 时，表示 AP 对应射频上的 VAP 已创建成功。

```
<AC>display vap ssid Sales
Info: This operation may take a few seconds, please wait.
WID : WLAN ID
--------------------------------------------------------------------------------
AP ID  AP name  RfID  WID  BSSID           Status  Auth type      STA   SSID
--------------------------------------------------------------------------------
0      area_1   0     1    00E0-FCC5-07E0  ON      WPA/WPA2-PSK   0     Sales
0      area_1   1     1    00E0-FCC5-07E0  ON      WPA/WPA2-PSK   0     Sales
1      area_2   0     1    00E0-FC03-0490  ON      WPA/WPA2-PSK   0     Sales
1      area_2   1     1    00E0-FC03-04A0  ON      WPA/WPA2-PSK   0     Sales
--------------------------------------------------------------------------------
Total: 4
```

（2）将 STA1 接入 Sales，STA2 和 Phone1 接入 Finances 的无线网络后，在无线控制器 AC 上使用 "display station ssid Sales" 和 "display station ssid Finances" 命令，查看已接入无线网络中的用户。

```
<AC>display station ssid Sales
Rf/WLAN: Radio ID/WLAN ID
Rx/Tx: link receive rate/link transmit rate(Mbps)
```

```
--------------------------------------------------------------------------------
STA MAC         AP ID Ap name   Rf/WLAN  Band  Type   Rx/Tx    RSSI  VLAN  IP address
--------------------------------------------------------------------------------
5489-988e-18dc   0    area_1    0/2      2.4G   -      -/-      -     11    10.0.11.29
--------------------------------------------------------------------------------
Total: 1 2.4G: 1 5G: 0
<AC>display station ssid Finances
Rf/WLAN: Radio ID/WLAN ID
Rx/Tx: link receive rate/link transmit rate(Mbps)
--------------------------------------------------------------------------------
STA MAC         AP ID Ap name   Rf/WLAN  Band  Type   Rx/Tx    RSSI  VLAN  IP address
--------------------------------------------------------------------------------
5489-9808-362a   1    area_2    0/2      2.4G   -      -/-      -     12    10.0.102.69
5489-9880-5d4d   0    area_1    0/2      2.4G   -      -/-      -     12    10.0.102.183
--------------------------------------------------------------------------------
Total: 2 2.4G: 2 5G: 0
```

（3）测试 STA 获取 IP 地址及漫游情况。在 STA1 和 STA2 正常关联无线网络后，使用"ipconfig"命令查看 STA1 和 STA2 通过无线网络自动获取的 IP 地址。

（4）测试 STA1 与其他站点的网络连通性，在连通性测试结果中可以看到全网互通。

任务 21　实现网络设备的远程管理

任务目标

1．了解远程管理的作用。
2．实现网络设备的远程管理。

任务描述

某公司的网络管理员小赵负责公司办公网的管理工作，熟悉公司内部设备的运行情况，每天都需要保障公司内部网络设备的正常运行，同时进行办公网的日常管理和维护工作。

在安装的办公网中，路由器和交换机放置在中心机房，每次都需要去中心机房进行现场配置、调试，非常麻烦。因此小赵决定在路由和交换设备上开启远程登录管理功能，即通过远程方式登录路由器和交换机。

任务要求

（1）实现网络设备的远程管理功能，网络拓扑结构如图 21.1 所示。

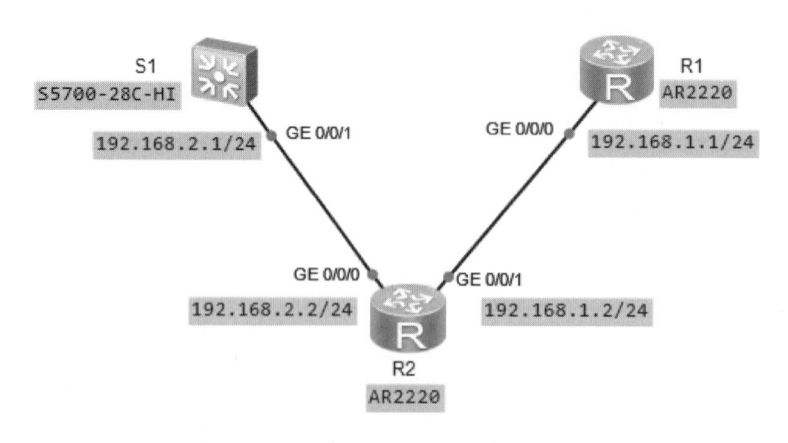

图 21.1　实现网络设备的远程管理的网络拓扑结构

（2）各路由器和交换机端口及 IP 地址设置如表 21.1 所示。

表 21.1　各路由器和交换机的端口及 IP 地址设置

设 备 名 称	端　　口	IP 地址/子网掩码
R1	GE 0/0/0	192.168.1.1/24
S1	GE 0/0/1（VLANIF1）	192.168.2.1/24
R2	GE 0/0/0	192.168.2.2/24
	GE 0/0/1	192.168.1.2/24

（3）在路由器 R1 和交换机 S1 上配置 STelnet 远程管理，并在路由器 R2 上登录路由器 R1 进行验证。

任务实施

步骤 1：按照图 21.1 搭建网络拓扑结构，连线全部使用直通线，开启所有设备电源。

步骤 2：对路由器 R1 进行基本配置。

```
<Huawei>system-view
[Huawei]sysname R1
[R1]interface GigabitEthernet 0/0/0
[R1-GigabitEthernet0/0/0]ip address 192.168.1.1 24
[R1-GigabitEthernet0/0/0]quit
```

步骤 3：对交换机 S1 进行基本配置。

```
<Huawei>system-view
[Huawei]sysname S1
[S1]int Vlanif 1
[S1-Vlanif1]ip add 192.168.2.1 24
[S1-Vlanif1]quit
```

步骤 4：对路由器 R2 进行基本配置。

```
<Huawei>system-view
[Huawei]sysname R2
[R2]interface GigabitEthernet 0/0/0
[R2-GigabitEthernet0/0/0]ip address 192.168.2.2 24
[R2-GigabitEthernet0/0/0]quit
[R2]interface GigabitEthernet 0/0/1
[R2-GigabitEthernet0/0/1]ip address 192.168.1.2 24
```

步骤 5：使用 "rsa local-key-pair create" 命令生成本地 RSA 密钥对。

完成 SSH 登录的首要操作是配置并生成本地 RSA 密钥对。在进行其他 SSH 配置之前要先生成本地 RSA 密钥对，并将生成的本地 RSA 密钥对保存在设备中，重启后不会丢失。

（1）在路由器 R1 上生成本地 RSA 密钥对。

```
[R1]rsa local-key-pair create
The key name will be: Host
```

```
% RSA keys defined for Host already exist.
Confirm to replace them? (y/n)[n]:y
The range of public key size is (512 ~ 2048).
NOTES: If the key modulus is greater than 512,
       It will take a few minutes.
Input the bits in the modulus[default = 512]:512
Generating keys...
......++++++++++++
.......++++++++++++
......++++++++
.++++++++
```

(2) 在交换机 S1 上生成本地 RSA 密钥对。

```
[S1]rsa local-key-pair create
The key name will be: S1_Host
% RSA keys defined for Huawei_Host already exist.
Confirm to replace them? [y/n]:y
The range of public key size is (512 ~ 2048).
NOTES: If the key modulus is greater than 512,
       it will take a few minutes.
Input the bits in the modulus[default = 512]:
Generating keys...
....++++++++++++
.............+++++++++++
...++++++++
.............++++++++
```

步骤 6：开启 SSH 服务，配置 SSH 用户登录界面，并配置用户验证方式为 AAA 授权验证，用户名为"admin"，密码为"Huawei"。

(1) 在路由器 R1 上进行配置。

```
[R1]stelnet server enable
[R1]user-interface vty 0 4                      //进入 vty 用户界面
[R1-ui-vty0-4]authentication-mode aaa           //配置用户验证方式为 AAA
[R1-ui-vty0-4]user privilege level 2            //配置本地用户的优先级
[R1-ui-vty0-4]protocol inbound ssh              //只支持 SSH 协议，禁止 Telnet 功能
[R1-ui-vty0-4]idle-timeout 15                   //断连时间为 15 分钟
[R1-ui-vty0-4]quit
[R1]aaa                                         //进入 aaa 视图
[R1-aaa]local-user admin password cipher Huawei //创建本地用户和口令，以密文方式显示用户口令
[R1-aaa]local-user admin service-type ssh       //配置本地用户的接入类型为 SSH
[R1-aaa]quit
```

(2) 在交换机 S1 上进行配置。

```
[S1]stelnet server enable
[S1]ssh user admin authentication-type password    //配置用户 admin 的 SSH 认证为 password
```

```
[S1]ssh user admin service-type stelnet     //配置用户admin的SSH服务类型为STelnet
[S1]aaa
[S1-aaa]local-user admin password cipher Huawei
[S1-aaa]local-user admin service-type ssh
[S1-aaa]quit
[S1]user-interface vty 0 4
[S1-ui-vty0-4]authentication-mode aaa
[S1-ui-vty0-4]user privilege level 2
[S1-ui-vty0-4]protocol inbound ssh
[S1-ui-vty0-4]idle-timeout 15
[S1-ui-vty0-4]quit
```

步骤7：配置SSH客户端首次认证功能。

当用户第一次登录SSH服务器时，SSH客户端还没有保存SSH服务器的RSA公钥，因此对SSH服务器的RSA公钥的有效性检查失败，从而导致登录SSH服务器失败。因此当首次登录客户端的路由器R2时，需要开启SSH客户端的首次认证功能，不对SSH服务器的RSA公钥进行有效性检查。

```
[R2]ssh client first-time enable
```

任务验收

（1）使用"stelnet 192.168.1.1"命令在路由器R2上登录路由器R1。

```
[R2]stelnet 192.168.1.1
Please input the username:admin
Trying 192.168.1.1 ...
Press CTRL+K to abort
Connected to 192.168.1.1 ...
The server is not authenticated. Continue to access it? (y/n)[n]:y
Jul 20 2021 14:06:37-08:00 R1 %%01SSH/4/CONTINUE_KEYEXCHANGE(l)[1]:The server had
not been authenticated in the process of exchanging keys. When deciding whether
to continue, the user chose Y.
[SW3A]
Save the server's public key? (y/n)[n]:y
The server's public key will be saved with the name 192.168.2.1. Please wait...

Jul 20 2021 14:06:40-08:00 R1 %%01SSH/4/SAVE_PUBLICKEY(l)[2]:When deciding whether
to save the server's public key 192.168.1.1, the user chose Y.
[SW3A]
Enter password:                   //输入密码为"Huawei"
<R1>system-view
Enter system view, return user view with Ctrl+Z.
[R1]
```

（2）使用"display users"命令查看已经登录的用户信息，以路由器 R1 为例。

```
[R1]display users
  User-Intf    Delay     Type    Network Address    AuthenStatus    AuthorcmdFlag
+ 0  CON 0    00:00:00                                  pass
  Username : Unspecified

  129 VTY 0   00:01:59   SSH     192.168.2.1            pass
  Username : admin
```

任务 22 使用基本 ACL 限制网络访问

任务目标

1．了解访问控制列表的工作原理和分类。
2．实现基本 ACL 的配置。

任务描述

某公司构建了互通的办公网，为保护公司内网用户数据的安全，该公司内网采取安全防范措施。该公司分为技术部、财务部和销售部，分属 3 个不同的网段，3 个部门的计算机之间用路由器进行信息传递。为了安全起见，该公司领导要求网络管理员小赵对网络的数据流量进行控制，使技术部的计算机不能访问销售部的网络，但可以访问财务部的网络。

任务要求

在路由器上应用标准的 ACL（Access Control Lists，访问控制列表），对销售部的网络访问进行限制，禁止技术部计算机访问销售部计算机的网络数据流量通过，但对财务部的网络访问不进行限制，从而达到保护销售部计算机安全的目的。

（1）使用基本 ACL 限制网络访问，网络拓扑结构如图 22.1 所示。

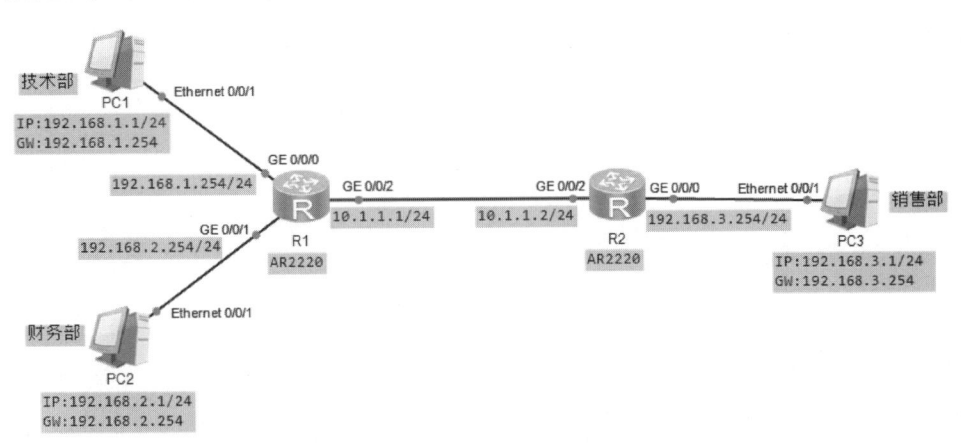

图 22.1 使用基本 ACL 限制网络访问的网络拓扑结构

（2）路由器的端口及 IP 地址设置如表 22.1 所示。

表 22.1 路由器的端口及 IP 地址设置

设 备 名 称	端 口	IP 地址/子网掩码
R1	GE 0/0/0	192.168.1.254/24
	GE 0/0/1	192.168.2.254/24
	GE 0/0/2	10.1.1.1/24
R2	GE 0/0/0	192.168.3.254/24
	GE 0/0/2	10.1.1.2/24

（3）PC1～PC3 的端口及 IP 地址设置如表 22.2 所示。

表 22.2 PC1～PC3 的端口及 IP 地址设置

设 备 名 称	端 口	IP 地址/子网掩码	网 关	备 注
PC1	Ethernet 0/0/1	192.168.1.1/24	192.168.1.254	技术部
PC2	Ethernet 0/0/1	192.168.2.1/24	192.168.2.254	财务部
PC3	Ethernet 0/0/1	192.168.3.1/24	192.168.3.254	销售部

（4）使用静态路由协议来实现全网互通。

（5）配置基本 ACL，限制 PC1 所在的网络可以访问 PC2 所在的网络，不能访问 PC3 所在的网络，但允许 PC2 所在的网络访问 PC3 所在的网络。

任务实施

步骤 1：按照图 22.1 搭建网络拓扑结构，连线全部使用直通线，开启所有设备电源，并为每台计算机设置相应的 IP 地址、子网掩码和默认网关。

步骤 2：对路由器 R1 进行基本配置。

```
<Huawei>system-view
[Huawei]sysname R1
[R1]interface GigabitEthernet 0/0/0
[R1-GigabitEthernet0/0/0]ip add 192.168.1.254 24
[R1-GigabitEthernet0/0/0]quit
[R1]interface GigabitEthernet 0/0/1
[R1-GigabitEthernet0/0/1]ip add 192.168.2.254 24
[R1-GigabitEthernet0/0/1]quit
```

```
[R1]interface GigabitEthernet0/0/2
[R1-GigabitEthernet0/0/2]ip add 10.1.1.1 24
[R1-GigabitEthernet0/0/2]quit
```

步骤 3：对路由器 R2 进行基本配置。

```
<Huawei>system-view
[Huawei]sysname R2
[R2]interface GigabitEthernet 0/0/0
[R2-GigabitEthernet0/0/0]ip add 192.168.3.254 24
[R2-GigabitEthernet0/0/0]quit
[R2]interface GigabitEthernet0/0/2
[R2-GigabitEthernet0/0/2]ip add 10.1.1.2 24
[R2-GigabitEthernet0/0/2]quit
```

步骤 4：配置静态路由实现全网互通。

（1）在路由器 R1 上进行配置。

```
[R1]ip route-static 192.168.3.0 255.255.255.0 10.1.1.2
```

（2）在路由器 R2 上进行配置。

```
[R2]ip route-static 192.168.1.0 255.255.255.0 10.1.1.1
[R2]ip route-static 192.168.2.0 255.255.255.0 10.1.1.1
```

步骤 5：配置基本 ACL，并查看 ACL 信息。

```
[R2]acl 2000
[R2-acl-basic-2000]rule deny source 192.168.1.0 0.0.0.255
[R2-acl-basic-2000]quit
[R2]display acl all
 Total quantity of nonempty ACL number is 1
Basic ACL 2000, 1 rule
Acl's step is 5
 rule 5 deny source 192.168.1.0 0.0.0.255
```

步骤 6：在端口上使用 ACL 限制网络访问。

```
[R2]interface GigabitEthernet 0/0/0
[R2-GigabitEthernet0/0/0]traffic-filter outbound acl 2000
[R2-GigabitEthernet0/0/0]quit
```

任务验收

（1）在 PC1 上测试其与 PC3 的连通性，通过结果可以看到二者是不连通的；在 PC1 上测试其与 PC2 的连通性，通过结果可以看到二者是连通的，如图 22.2 所示。

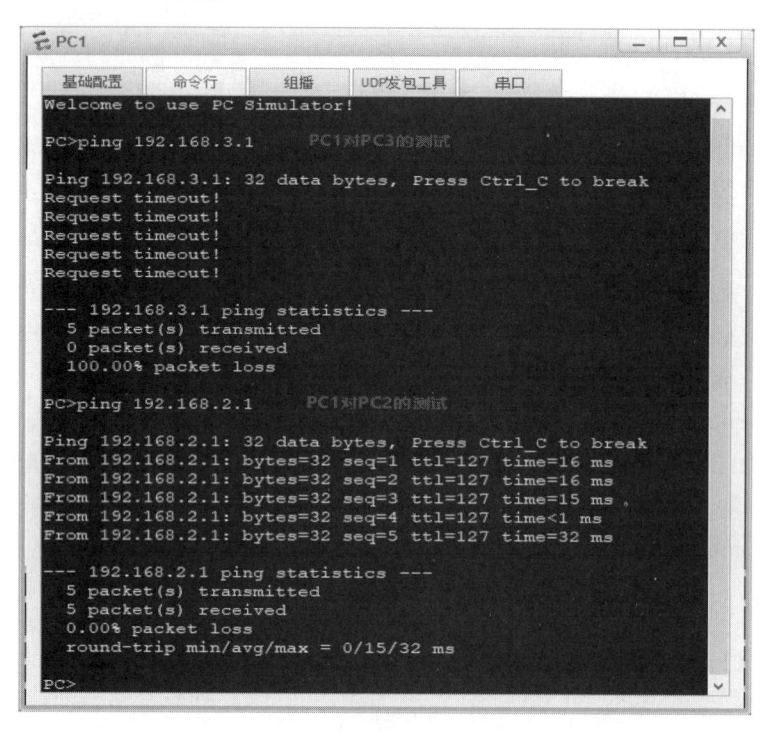

图 22.2　在 PC1 上测试其与 PC3 和 PC2 的连通性

（2）在 PC2 上测试其与 PC3 的连通性，如图 22.3 所示，通过结果可以看到二者是连通的。

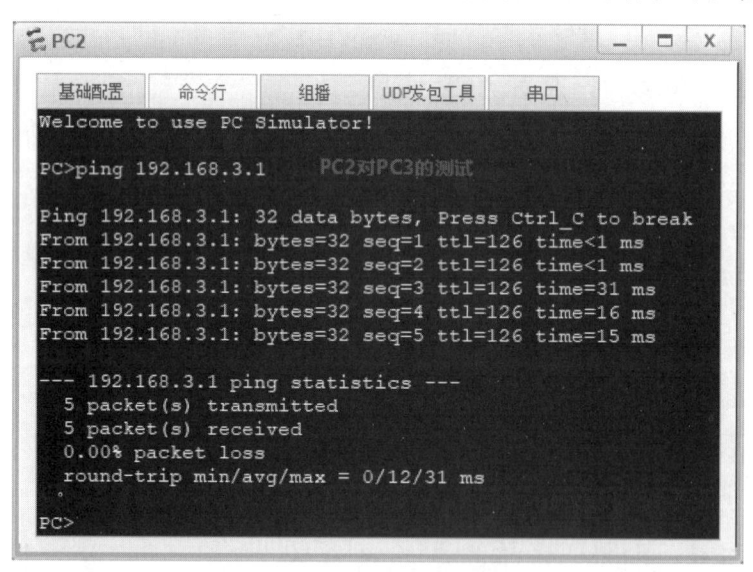

图 22.3　在 PC2 上测试其与 PC3 的连通性

（3）查看 ACL 的应用状态。

```
[R2]display acl 2000
Basic ACL 2000, 1 rule
Acl's step is 5
 rule 5 deny source 192.168.1.0 0.0.0.255 (5 matches)
```

任务 23　使用高级 ACL 限制服务器端口

任务目标

1. 了解基本 ACL 与高级 ACL 的区别。
2. 实现高级 ACL 的配置。

任务描述

某公司在天津设有分公司，其使用三层设备的专线技术，借助互联网和总公司网络实现连通。由于天津分公司的网络安全措施不严密，该公司规定禁止天津分公司销售部的计算机访问总公司的 FTP 服务器资源，允许天津分公司销售部和技术部的计算机访问总公司的 Web 等公开信息资源，而对服务器的其他访问均被限制，从而达到保护服务器和数据安全的目的。

任务要求

（1）使用高级 ACL 限制服务器端口，网络拓扑结构如图 23.1 所示。

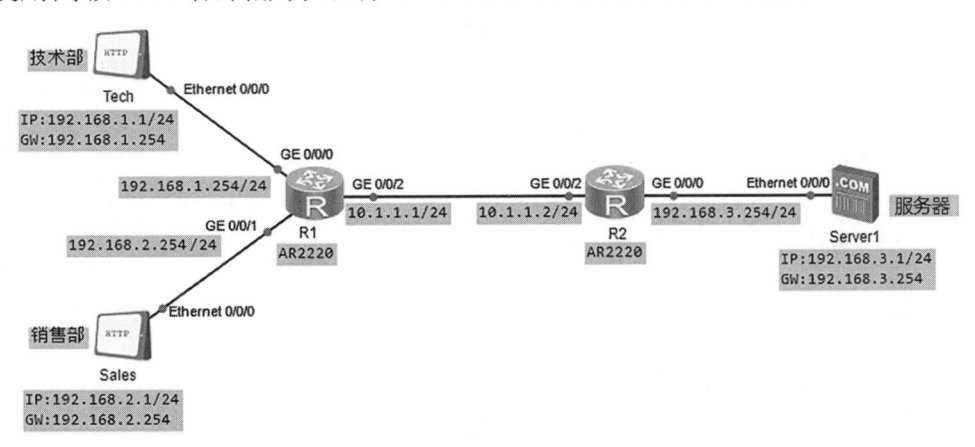

图 23.1　使用高级 ACL 限制服务器端口的网络拓扑结构

（2）路由器的端口及 IP 地址设置如表 23.2 所示。

表 23.2 路由器的端口及 IP 地址设置

设备名称	端口	IP 地址/子网掩码
R1	GE 0/0/0	192.168.1.254/24
	GE 0/0/1	192.168.2.254/24
	GE 0/0/2	10.1.1.1/24
R2	GE 0/0/0	192.168.3.254/24
	GE 0/0/2	10.1.1.2/24

（3）计算机的端口及 IP 地址参数如表 23.3 所示。

表 23.3 计算机的端口及 IP 地址参数

设备名称	端口	IP 地址/子网掩码	网关	备注
Tech	Ethernet 0/0/0	192.168.1.1/24	192.168.1.254	技术部
Sales	Ethernet 0/0/0	192.168.2.1/24	192.168.2.254	销售部
Server1	Ethernet 0/0/0	192.168.3.1/24	192.168.3.254	服务器

（4）使用静态路由实现全网互通。

（5）配置高级 ACL，限制 Sales 访问总公司的 FTP 服务器资源，允许 Sales 和 Tech 访问总公司的 Web 等公开信息资源。

任务实施

步骤 1：按照图 23.1 搭建网络拓扑图，连线全部使用直通线，开启所有设备电源，并为每台计算机设置相应的 IP 地址、子网掩码和默认网关。

步骤 2：对路由器 R1 进行基本配置。

```
<Huawei>system-view
[Huawei]sysname R1
[R1]interface GigabitEthernet 0/0/0
[R1-GigabitEthernet0/0/0]ip add 192.168.1.254 24
[R1-GigabitEthernet0/0/0]quit
[R1]interface GigabitEthernet 0/0/1
[R1-GigabitEthernet0/0/1]ip add 192.168.2.254 24
[R1-GigabitEthernet0/0/1]quit
[R1]interface GigabitEthernet0/0/2
[R1-GigabitEthernet0/0/2]ip add 10.1.1.1 24
[R1-GigabitEthernet0/0/2]quit
```

步骤 3：对路由器 R2 进行基本配置。

```
<Huawei>system-view
[Huawei]sysname R2
```

```
[R2]interface GigabitEthernet 0/0/0
[R2-GigabitEthernet0/0/0]ip add 192.168.3.254 24
[R2-GigabitEthernet0/0/0]quit
[R2]interface GigabitEthernet0/0/2
[R2-GigabitEthernet0/0/2]ip add 10.1.1.2 24
[R2-GigabitEthernet0/0/2]quit
```

步骤 4：配置静态路由实现全网互通。

（1）在路由器 R1 上进行配置。

```
[R1]ip route-static 192.168.3.0 255.255.255.0 10.1.1.2
```

（2）在路由器 R2 上进行配置。

```
[R2]ip route-static 192.168.1.0 255.255.255.0 10.1.1.1
[R2]ip route-static 192.168.2.0 255.255.255.0 10.1.1.1
```

步骤 5：配置高级 ACL。

```
[R1]acl 3000
[R1-acl-adv-3000]rule 5 deny tcp source 192.168.2.0 0.0.0.255 destination 192.168.3.1 0.0.0.0 destination-port range 20 21
[R1-acl-adv-3000]rule 10 permit tcp source 192.168.2.0 0.0.0.255 destination 192.168.3.1 0.0.0.0 destination-port eq 80
[R1-acl-adv-3000]rule 15 deny ip
[R1-acl-adv-3000]quit
```

步骤 6：查看 ACL 信息。

```
[R1]display acl all
 Total quantity of nonempty ACL number is 1
Advanced ACL 3000, 3 rules
Acl's step is 5
 rule 5 deny tcp source 192.168.2.0 0.0.0.255 destination 192.168.3.1 0 destination-port range ftp-data ftp
 rule 10 permit tcp source 192.168.2.0 0.0.0.255 destination 192.168.3.1 0 destination-port eq www
 rule 15 deny ip
```

步骤 7：在端口上使用 ACL 限制网络访问。

```
[R1]interface GigabitEthernet 0/0/1
[R1-GigabitEthernet 0/0/1]traffic-filter inbound acl 3000
[R1-GigabitEthernet 0/0/1]quit
```

步骤 8：配置 Server1 的 FtpServer 和 HttpServer。

（1）配置 FtpServer。

先右击 Server1，在弹出的快捷菜单中选择"设置"命令，再在打开的设置对话框的"服务器信息"选项卡中选择"FtpServer"选项，在"配置"选区中进行文件根目录的配置，这里选择"C:\"，如图 23.2 所示，最后单击"启动"按钮。

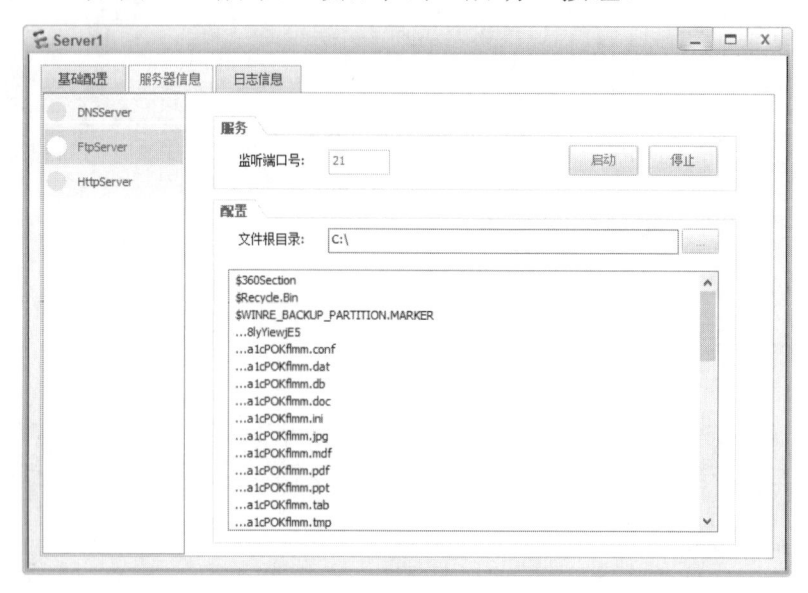

图 23.2　配置 FtpServer

（2）配置 HttpServer。

先右击 Server1，在弹出的快捷菜单中选择"设置"命令，再在打开的设置对话框的"服务器信息"选项卡中选择"HttpServer"选项，在"配置"选区中进行文件根目录的配置，这里选择"C:\Http\index.htm"（需提前创建好），如图 23.3 所示，最后单击"启动"按钮。

图 23.3　配置 HttpServer

任务验收

（1）Sales 可以正常访问 Web 服务器，如图 23.4 所示。

图 23.4　Sales 正常访问 Web 服务器

（2）Sales 无法正常访问 FTP 服务器，如图 23.5 所示。

图 23.5　Sales 无法正常访问 FTP 服务器

（3）查看高级 ACL 应用状态。

```
[R1]dispaly acl all
 Total quantity of nonempty ACL number is 1
 Advanced ACL 3000, 3 rules
 Acl's step is 5
  rule 5 deny tcp source 192.168.2.0 0.0.0.255 destination 192.168.3.1 0 destination-port range ftp-data ftp (5 matches)
  rule 10 permit tcp source 192.168.2.0 0.0.0.255 destination 192.168.3.1 0 destination-port eq www (6 matches)
  rule 15 deny ip
```

任务 24　实现公司内网安全接入互联网

任务目标

1. 了解 PAP 与 CHAP 认证的基本过程。
2. 实现 PPP 封装 PAP 和 CHAP 认证的配置和认证方法。

任务描述

某公司借助专线接入技术，将总部网络接入互联网，利用互联网与该公司位于天津的分公司网络中心的路由器连接，从而实现公司网络连通。

任务要求

为了保护公司总部网络和分公司的网络安全，对公司网络中的路由器设置 PPP（Point-to-Point Protocol，点到点协议）安全认证，客户端路由器与电信运营商在进行链路协商时要认证身份，以实现公司全网安全通信。

（1）实现公司内网安全接入互联网，网络拓扑结构如图 24.1 所示。

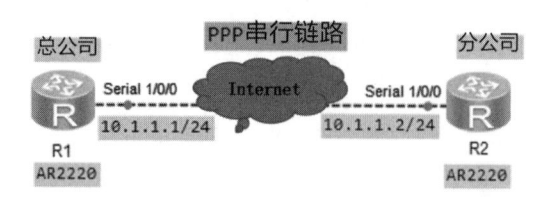

图 24.1　实现公司内网安全接入互联网的网络拓扑结构

（2）路由器的端口及 IP 地址设置如表 24.1 所示。

表 24.1　路由器的端口及 IP 地址设置

设备名称	端口	IP 地址/子网掩码
R1	Serial 1/0/0	10.1.1.1/24
R2	Serial 1/0/0	10.1.1.2/24

（3）在两台路由器之间设置 PPP 封装，并测试两台路由器的连通性。

任务实施

步骤 1：PPP 封装 PAP 安全认证。

在两台路由器之间设置 PPP 封装 PAP（Password Authentication Protocol，密码认证协议）认证，R1 作为认证方可以配置认证的用户名和密码，并确定该用户名和密码用于 PAP 认证；R2 作为被认证方可以配置以 PAP 认证时本地发送的 PAP 用户名 "admin" 和密码 "huawei"，并测试两台路由器的连通性。

（1）按照图 24.1 搭建网络拓扑，为两台路由器添加 2SA 模块，并且添加在 Serial 1/0/0 端口位置；连线使用 Serial 串口线，并开启所有设备电源。

（2）对路由器 R1 进行基本配置。

```
<Huawei>system-view
[Huawei]sysname R1
[R1]interface Serial 1/0/0
[R1-Serial1/0/0]ip add 10.1.1.1 24
[R1-Serial1/0/0]quit
```

（3）对路由器 R2 进行基本配置。

```
<Huawei>system-view
[Huawei]sysname R2
[R2]interface Serial 1/0/0
[R2-Serial1/0/0]ip add 10.1.1.2 24
[R2-Serial1/0/0]quit
```

（4）配置 PPP 的 PAP 认证。

路由器 R1 作为认证方，需要配置本端 PPP 的认证方式为 PAP。执行 "aaa" 命令，进入 AAA 视图，配置 PAP 认证所使用的用户名和密码。

```
[R1]aaa
[R1-aaa]local-user amin password cipher Huawei  //在路由器 R1 上指定该密码用于 PPP 认证
[R1-aaa]local-user admin service-type ppp
[R1-aaa]interface Serial 1/0/0
[R1-Serial1/0/0]link-protocol ppp
[R1-Serial1/0/0]ppp authentication-mode pap
//在 Serial 1/0/0 端口上启用 PPP 认证，并指定认证方式为 PAP
[R1-Serial1/0/0]quit
```

（5）查看路由器 R1 的链路状态信息。

关闭路由器 R1 与路由器 R2 相连的端口一段时间后打开，使路由器 R1 与路由器 R2 之间的链路重新协商，并检查链路状态和连通性。

```
[R1]interface Serial 1/0/0
```

```
[R1-Serial1/0/0]shutdown
[R1-Serial1/0/0]undo shutdown
[R1-Serial1/0/0]quit
[R1]display ip int brief
Interface                IP Address/Mask         Physical        Protocol
GigabitEthernet0/0/0     unassigned              down            down
GigabitEthernet0/0/1     unassigned              down            down
GigabitEthernet0/0/2     unassigned              down            down
NULL0                    unassigned              up              up(s)
Serial1/0/0              10.1.1.1/24             up              down
Serial1/0/1              unassigned              down            down
//可以观察到，现在路由器 R1 和路由器 R2 之间无法正常通信，链路物理状态正常，但是链路层协议状态不正常，这是因为此时 PPP 链路上的 PAP 认证未通过
```

（6）配置对端（被认证方）PAP 认证。

路由器 R2 作为被认证端，在 Serial 1/0/0 端口中配置以 PAP 认证时本地发送的 PAP 用户名和密码。

```
[R2]int Serial 1/0/0
[R2-Serial1/0/0]link-protocol ppp
[R2-Serial1/0/0]ppp pap local-user admin password cipher Huawei
//在 Serial 1/0/0 端口上启用 PPP，并指定 PAP 认证的用户名和密码
```

（7）查看路由器 R1 的链路状态信息。

```
[R1]dis ip int brief
Interface                IP Address/Mask         Physical        Protocol
GigabitEthernet0/0/0     unassigned              down            down
GigabitEthernet0/0/1     unassigned              down            down
GigabitEthernet0/0/2     unassigned              down            down
NULL0                    unassigned              up              up(s)
Serial1/0/0              10.10.10.1/24           up              up
Serial1/0/1              unassigned              down            down
//可以观察到，现在路由器 R1 与路由器 R2 之间的链路层协议的状态正常
```

步骤 2：PPP 封装 CHAP 安全认证。

在两台路由器之间设置 PPP 封装 CHAP（Challenge Handshake Authentication Protocol，挑战握手身份认证协议）认证，R1 作为认证方，需要配置本端 PPP 协议的认证方式为 CHAP；R2 作为被认证方，配置以 CHAP 方式认证时本地发送的 CHAP 用户名"admin"和密码"huawei"，并测试两台路由器的连通性。

（1）按照图 24.1 搭建网络拓扑结构，在两台路由器的 Serial 1/0/0 端口位置上分别添加 2SA 模块，连线使用 Serial 串口线，并开启所有设备电源。

（2）对路由器 R1 进行基本配置。

```
<Huawei>system-view
```

```
[Huawei]sysname R1
[R1]int Serial 1/0/0
[R1-Serial1/0/0]ip add 10.1.1.1 24
[R1-Serial1/0/0]quit
```

（3）对路由器 R2 进行基本配置。

```
<Huawei>system-view
[Huawei]sysname R2
[R2]int Serial 1/0/0
[R2-Serial1/0/0]ip add 10.1.1.2 24
[R2-Serial1/0/0]quit
```

（4）配置 PPP 协议的 CHAP 认证。

路由器 R1 作为认证端，需要配置本端 PPP 协议的认证方式为 CHAP。执行"aaa"命令，进入 AAA 视图，配置 CHAP 认证所使用的用户密码[①]。

```
[R1]aaa
[R1-aaa]local-user amin password cipher Huawei  //在路由器R1上指定密码应用于PPP认证
[R1-aaa]local-user admin service-type ppp
[R1-aaa]int s1/0/0
[R1-Serial1/0/0]link-protocol ppp
[R1-Serial1/0/0]ppp authentication-mode chap    //在Serial 1/0/0端口上启动PPP功能，并指
定认证方式为CHAP
[R1-Serial1/0/0]quit
```

（5）查看路由器 R1 的链路状态信息。

先关闭路由器 R1 与路由器 R2 相连的端口一段时间后再打开，使路由器 R1 与路由器 R2 之间的链路重新协商，并检查链路状态和连通性。

```
[R1]interface Serial 1/0/0
[R1-Serial1/0/0]shutdown
[R1-Serial1/0/0]undo shutdown
[R1]
[R1]display ip int brief
Interface                 IP Address/Mask      Physical      Protocol
GigabitEthernet0/0/0      unassigned           down          down
GigabitEthernet0/0/1      unassigned           down          down
GigabitEthernet0/0/2      unassigned           down          down
NULL0                     unassigned           up            up(s)
Serial1/0/0               10.1.1.1/24          up            down
Serial1/0/1               unassigned           down          down
//可以观察到，现在路由器R1和路由器R2之间无法正常通信，链路物理状态正常，但是链路层协议的状态不正
常，这是因为此时PPP链路上的PAP认证未通过
```

---

① 在本书中，s 表示 Serial。

（6）配置对端（被认证方）CHAP 认证。

路由器 R2 作为被认证端，在 Serial 1/0/0 端口上配置以 CHAP 方式认证时本地发送的 CHAP 用户名和密码。

```
[R2]int Serial 1/0/0
[R2-Serial1/0/0]link-protocol ppp
[R2-Serial1/0/0]ppp chap user admin
[R2-Serial1/0/0]ppp chap password cipher Huawei
//在 Serial 1/0/0 端口上启用 PPP 功能，并指定 CHAP 认证的用户名和密码
```

（7）查看路由器 R1 的链路状态信息。

```
[R1]display ip int brief
Interface                IP Address/Mask    Physical    Protocol
GigabitEthernet0/0/0     unassigned         down        down
GigabitEthernet0/0/1     unassigned         down        down
GigabitEthernet0/0/2     unassigned         down        down
NULL0                    unassigned         up          up(s)
Serial1/0/0              10.1.1.1/24        up          up
Serial1/0/1              unassigned         down        down
//可以观察到，现在路由器 R1 与路由器 R2 之间的链路层协议状态正常
```

任务验收

（1）在路由器 R2 上查看链路状态。

（2）在路由器 R1 上 ping 路由器 R2，测试路由器之间的连通性，通过结果可以发现二者是连通的。

```
[R1]ping 10.1.1.2
 PING 10.1.1.2: 56  data bytes, press CTRL_C to break
  Reply from 10.1.1.2: bytes=56 Sequence=1 ttl=255 time=30 ms
  Reply from 10.1.1.2: bytes=56 Sequence=2 ttl=255 time=20 ms
  Reply from 10.1.1.2: bytes=56 Sequence=3 ttl=255 time=20 ms
  Reply from 10.1.1.2: bytes=56 Sequence=4 ttl=255 time=40 ms
  Reply from 10.1.1.2: bytes=56 Sequence=5 ttl=255 time=20 ms
 --- 10.1.1.2 ping statistics ---
  5 packet(s) transmitted
  5 packet(s) received
  0.00% packet loss
  round-trip min/avg/max = 20/26/40 ms
```

任务 25　利用静态 NAT 技术实现外网计算机访问内网服务器

任务目标

1. 了解网络地址转换的原理和作用。
2. 理解网络地址转换的分类。
3. 掌握利用静态 NAT 技术实现外网计算机访问内网服务器的配置。

任务描述

某公司的办公网接入了互联网，由于需要进行企业宣传，需要建立用于产品推广和业务交流的网站。要求公司内部的 Web 服务器对外提供服务，因此在互联网上的计算机可以访问公司的内部网站。

目前，该公司只向网络运营商申请了 2 个公有 IP 地址。外网中的计算机是不能直接访问该公司内网服务器的，如果想要内网服务器中的服务能够被外网计算机访问，则需要将内网服务器的私有 IP 地址通过静态转换映射到公有 IP 地址上，只有这样互联网上的计算机才能通过公有 IP 地址访问内网服务器。

任务要求

（1）利用静态 NAT（Network Address Translation，网络地址转换）技术实现外网计算机访问内网服务器，网络拓扑结构如图 25.1 所示。

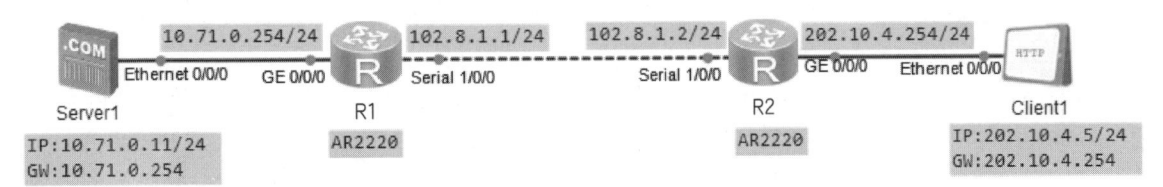

图 25.1　利用静态 NAT 技术实现外网计算机访问内网服务器的网络拓扑结构

（2）各路由器的端口及IP地址设置如表25.1所示。

表25.1 各路由器的端口及IP地址设置

设备名称	端口	IP地址/子网掩码
R1	GE 0/0/0	10.71.0.254/24
	Serial 1/0/0	102.8.1.1/24
R2	Serial 1/0/0	102.8.1.2/24
	GE 0/0/0	202.10.4.254/24

（3）客户端和服务器的端口及IP地址设置如表25.2所示。

表25.2 客户端和服务器的端口及IP地址设置

设备名称	本端端口	IP地址/子网掩码	默认网关	对端设备与端口
Client1	Ethernet 0/0/0	202.10.4.5/24	202.10.4.254	R1:GE 0/0/0
Server1	Ethernet 0/0/0	10.71.0.11/24	10.71.0.254	R2:GE 0/0/0

（4）在路由器R1上进行静态NAT配置，实现外网计算机能访问内网的Web服务器，映射地址为102.8.1.8。

任务实施

步骤1：按照图25.1搭建网络拓扑结构，在路由器的Serial 1/0/0端口上添加2SA模块，路由器之间的连线使用Serial串口线，其他使用直通线，开启所有设备电源，并为每台计算机设置相应的IP地址、子网掩码和默认网关。

步骤2：对路由器R1进行基本配置。

```
<Huawei>system-view
[Huawei]sysname R1
[R1]interface GigabitEthernet 0/0/0
[R1-GigabitEthernet0/0/0]ip add 10.71.0.254 24
[R1-GigabitEthernet0/0/0]quit
[R1]interface Serial 1/0/0
[R1-Serial1/0/0]ip add 102.8.1.1 24
[R1-Serial1/0/0]quit
```

步骤3：对路由器R2进行基本配置。

```
<Huawei>system-view
[Huawei]sysname R2
[R2]interface GigabitEthernet 0/0/0
[R2-GigabitEthernet0/0/0]ip add 202.10.4.254 24
[R2-GigabitEthernet0/0/0]quit
[R2]interface Serial 1/0/0
[R2-Serial1/0/0]ip add 102.8.1.2 24
```

```
[R2-Serial1/0/0]quit
```

步骤 4：在路由器 R1 上配置默认路由。

```
[R1]ip route-static 0.0.0.0 0.0.0.0  Serial1/0/0
```

步骤 5：在路由器 R1 上配置静态 NAT 映射。

```
[R1]int s1/0/0
[R1-Serial1/0/0]nat static global 102.8.1.8 inside 10.71.0.11
//将地址 102.8.1.8 映射到私网地址 10.71.0.11
```

步骤 6：在 Server1 上配置 HttpServer。

先右击 Server1，在弹出的快捷菜单中选择"设置"命令，再在打开的设置对话框的"服务器信息"选项卡中选择"HttpServer"选项，在"配置"选区中进行文件根目录的设置，这里选择"C:\Http\index.htm"（需要提前创建好，网页文件的内容可以为空），如图 25.2 所示，最后单击"启动"按钮。

图 25.2 配置 HttpServer

任务验收

（1）在 Client1 上尝试访问 http://102.8.1.8，如图 25.3 所示，发现可以正常访问公司内网的 Web 服务器。

图 25.3　Client1 访问公司内网的 Web 服务器

（2）在路由器 R1 上查看静态 NAT 映射关系。

```
[R1]display nat static
 Static Nat Information:
 Interface : Serial1/0/0
   Global IP/Port    : 102.8.1.8/----
   Inside IP/Port    : 10.71.0.11/----
   Protocol : ----
   VPN instance-name : ----
   Acl number        : ----
   Netmask : 255.255.255.255
   Description : ----
 Total :    1
```
//从以上显示信息中可以看出，已经成功地在路由器 R1 上配置了静态 NAT 映射，实现了 Web 服务器的私有 IP 地址 10.71.0.11 与公有 IP 地址 102.8.1.8 的映射

任务 26　利用动态 NAPT 技术实现局域网计算机访问互联网

任务目标

1．了解网络地址转换的原理和作用。
2．理解网络地址转换的分类。
3．掌握利用动态 NAPT 技术实现局域网计算机访问互联网的配置。

任务描述

由于业务的需要，某公司的办公网需要接入互联网，网络管理员小赵向网络运营商申请了一条专线，该专线分配了 4 个公有 IP 地址，要求公司所有部门的计算机都能访问互联网。

公司通过路由器接入互联网，采用动态 NAPT（Network Address Port Translation，网络端口地址转换）技术，实现局域网的多台计算机共用一个或少数几个公有 IP 地址访问互联网。

任务要求

（1）利用动态 NAPT 技术实现局域网计算机访问互联网，网络拓扑结构如图 26.1 所示。

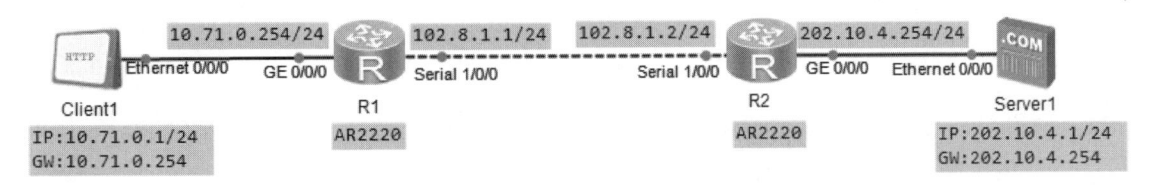

图 26.1　利用动态 NAPT 技术实现局域网计算机访问互联网的网络拓扑结构

（2）路由器的端口及 IP 地址设置如表 26.1 所示。

表 26.1 路由器的端口及 IP 地址设置

设 备 名 称	端 口	IP 地址/子网掩码
R1	GE 0/0/0	10.71.0.254/24
	Serial 1/0/0	102.8.1.1/24
R2	Serial 1/0/0	102.8.1.2/24
	GE 0/0/0	202.10.4.254/24

（3）客户端和服务器的端口及 IP 地址设置如表 26.2 所示。

表 26.2 客户端和服务器的端口及 IP 地址设置

设 备 名 称	本 端 端 口	IP 地址/子网掩码	网 关	对端设备与端口
Client1	Ethernet 0/0/0	10.71.0.1/24	10.71.0.254	R1:GE 0/0/0
Server1	Ethernet 0/0/0	202.10.4.1/24	202.10.4.254	R2:GE 0/0/0

（4）在路由器 R1 上进行动态 NAPT 配置，局域网计算机能通过公有 IP 地址访问互联网上的服务器，动态 NAPT 地址池使用 IP 地址 102.8.1.3～102.8.1.5。

任务实施

步骤 1：按照图 26.1 搭建网络拓扑结构，在路由器的 Serial 1/0/0 端口上添加 2SA 模块，路由器之间的连线使用 Serial 串口线，其他使用直通线，开启所有设备电源，并为每台计算机设置相应的 IP 地址、子网掩码和默认网关。

步骤 2：对路由器 R1 进行基本配置。

```
<Huawei>system-view
[Huawei]sysname R1
[R1]interface GigabitEthernet 0/0/0
[R1-GigabitEthernet0/0/0]ip address 10.71.0.254 24
[R1-GigabitEthernet0/0/0]quit
[R1]interface Serial 1/0/0
[R1-Serial1/0/0]ip address 102.8.1.1 24
[R1-Serial1/0/0]quit
```

步骤 3：对路由器 R2 进行基本配置。

```
<Huawei>system-view
[Huawei]sysname R2
[R2]interface GigabitEthernet 0/0/0
[R2-GigabitEthernet0/0/0]ip address 202.10.4.254 24
[R2-GigabitEthernet0/0/0]quit
[R2]interface Serial 1/0/0
[R2-Serial1/0/0]ip address 102.8.1.2 24
[R2-Serial1/0/0]quit
```

步骤 4：在路由器 R1 上配置默认路由并验证。

```
[R1]ip route-static 0.0.0.0 0.0.0.0  Serial 1/0/0
[R1]ping 202.10.4.1
 PING 202.10.4.1: 56  data bytes, press CTRL_C to break
  Request time out
  Reply from 202.10.4.1: bytes=56 Sequence=2 ttl=254 time=20 ms
  Reply from 202.10.4.1: bytes=56 Sequence=3 ttl=254 time=30 ms
  Reply from 202.10.4.1: bytes=56 Sequence=4 ttl=254 time=40 ms
  Reply from 202.10.4.1: bytes=56 Sequence=5 ttl=254 time=20 ms
 --- 202.10.4.1 ping statistics ---
  5 packet(s) transmitted
  4 packet(s) received
  20.00% packet loss
  round-trip min/avg/max = 20/27/40 ms
```

步骤 5：在路由器 R1 上配置动态 NAPT。

```
[R1]nat address-group 1 102.8.1.3 102.8.1.5  //配置NAPT地址池
[R1]acl 2000
[R1-acl-basic-2000]rule permit source 10.71.0.0 0.0.0.255
[R1-acl-basic-2000]quit
[R1]int Serial 1/0/0
[R1-Serial1/0/0]nat outbound 2000 address-group 1
//用来配置动态 NAPT，将 ACL 和地址池关联起来，当命令行中没有 no-pat 参数时表示 NAPT
```

步骤 6：在 Server1 上配置 HttpServer。

先右击 Server1，在弹出的快捷菜单中选择"设置"命令，再在打开的设置对话框的"服务器信息"选项卡中选择"HttpServer"选项，在"配置"选区中进行文件根目录的设置，这里选择"C:\Http\index.htm"（需提前创建好，网页文件的内容可以为空），如图 26.2 所示，最后单击"启动"按钮。

图 26.2　配置 HttpServer

任务验收

（1）在 Client1 上尝试访问 http://202.10.4.1，如图 26.3 所示，通过结果可知能够正常访问 Web 服务器。

图 26.3 Client1 访问 Web 服务器

（2）在路由器 R1 上查看 NAPT 会话信息。

```
[R1]display nat session all
 NAT Session Table Information:
   Protocol          : TCP(6)
   SrcAddr  Port Vpn : 10.71.0.1        2312
   DestAddr Port Vpn : 202.10.4.1       20480
   NAT-Info
     New SrcAddr     : 102.8.1.4
     New SrcPort     : 10240
     New DestAddr    : ----
     New DestPort    : ----
 Total : 1
 //从以上显示信息中可以看出，Client1 的私有 IP 地址映射到了一个公有 IP 地址上
```

服务器部分

任务 27　安装与配置 Ubuntu 网络操作系统

任务目标

1. 了解 Linux 操作系统的组成及应用。
2. 安装 Ubuntu 网络操作系统。
3. 完成 Ubuntu 网络操作系统的基本配置。

任务描述

某公司购置了服务器，需要为服务器安装相应的操作系统。要求网络管理员小赵为新增的服务器安装与配置 Ubuntu 网络操作系统。

任务要求

对初学者来说，通过虚拟机软件来安装与配置 Ubuntu 网络操作系统是较好的选择。具体要求如下。

（1）准备"VMware Workstation 16 Pro"安装文件，可以在官网中下载。

（2）安装"VMware Workstation 16 Pro"应用程序。

（3）准备 Ubuntu 网络操作系统的 ISO 镜像文件（ubuntu-22.04.3-live-server-amd64.iso），可以在官网中下载。

（4）宿主机的 CPU 需要支持虚拟终端（Virtual Terminal，VT）技术，并处于开启状态。

（5）创建虚拟机和安装 Ubuntu 网络操作系统，项目参数如表 27.1 所示。

表 27.1　创建虚拟机和安装 Ubuntu 网络操作系统的项目参数

项目	条目	说明
创建虚拟机	类型	自定义（高级）
	客户机操作系统类型	Linux 的 64 位 Ubuntu
	虚拟机名称	Ubuntu-1

续表

项 目	条 目	说 明
创建虚拟机	存储位置	D:\Ubuntu-1
	网络类型	使用 NAT 模式
	内存大小	2GB
	硬盘类型和大小	SCSI，20GB
安装 Ubuntu 网络操作系统	安装过程中的语言	English
	键盘配置	English（US）
	网络连接	默认配置
	代理服务器和存储镜像配置	默认配置
	指导性存储配置、文件系统摘要	默认配置
	主机名	Ubuntu
	普通用户和密码	普通用户为"chris"，密码为"123456"
	OpenSSH 服务器	安装
	软件选择	默认配置
	其他项目	采用默认配置

（6）查看操作系统的相关信息。

（7）查看操作系统防火墙的相关信息。

任务实施

步骤 1：安装 VMware Workstation 16 Pro 应用程序。

关于 VMware Workstation 16 Pro 应用程序的安装方法，此处省略。

步骤 2：创建虚拟机。

（1）运行 VMware Workstation 16 Pro 应用程序，在主界面的菜单栏中选择"编辑"→"首选项"命令，如图 27.1 所示。

（2）在打开的"首选项"对话框中选择"工作区"选项，在对话框右侧区域中单击"浏览"按钮，或者在左侧的文本框中通过手动输入来设置虚拟机的默认位置，在本任务中将虚拟机的默认位置设置为"D:\"，如图 27.2 所示，设置完成后单击"确定"按钮。

（3）在 VMware Workstation 16 Pro 主界面中单击"创建新的虚拟机"按钮，如图 27.3 所示。

（4）在打开的"新建虚拟机向导"对话框中选择虚拟机的配置类型，其中"典型（推荐）"表示使用推荐的设置快速创建虚拟机，"自定义（高级）"表示根据需要设置虚拟机的硬件类型、兼容性、存储位置等。本任务选中"自定义（高级）"单选按钮，如图 27.4 所示，并单击"下一步"按钮。

图 27.1 VMware Workstation 16 Pro 应用程序主界面

图 27.2 设置虚拟机的默认位置

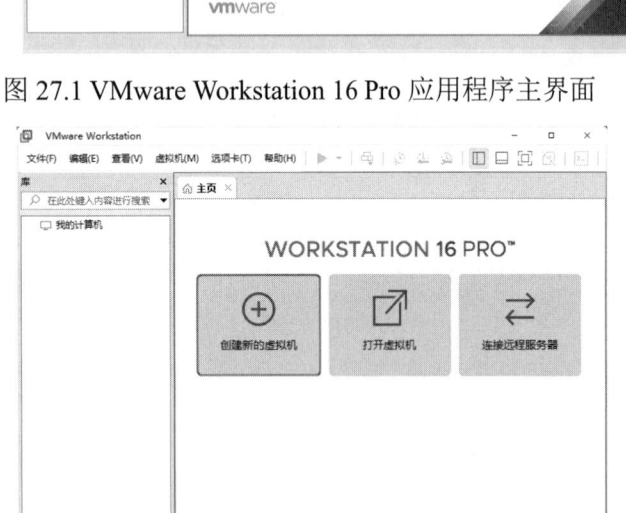

图 27.3 创建新的虚拟机

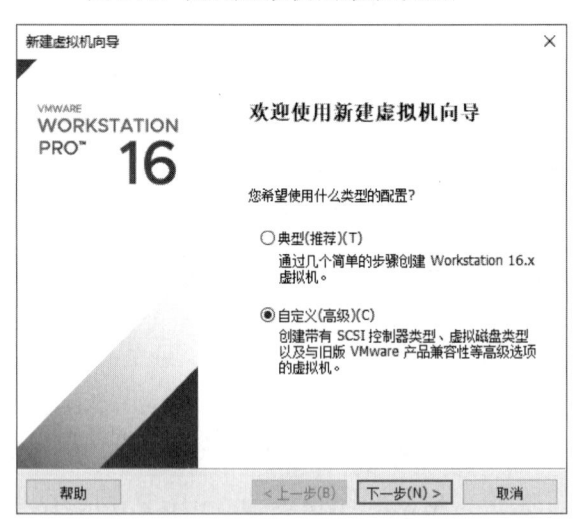

图 27.4 选择虚拟机的配置类型

（5）在打开的"选择虚拟机硬件兼容性"对话框中，单击"下一步"按钮。

（6）在打开的"安装客户机操作系统"对话框中选中"稍后安装操作系统"单选按钮，如图 27.5 所示，并单击"下一步"按钮。

（7）在打开的"选择客户机操作系统"对话框中，设置客户机操作系统为"Linux"，如图 27.6 所示，并单击"下一步"按钮。

（8）在打开的"命名虚拟机"对话框中输入虚拟机名称"Ubuntu-1"，如图 27.7 所示，并单击"下一步"按钮。

（9）在打开的"处理器配置"对话框中设置处理器数量及每个处理器的内核数量，并单击"下一步"按钮。

任务 27　安装与配置 Ubuntu 网络操作系统

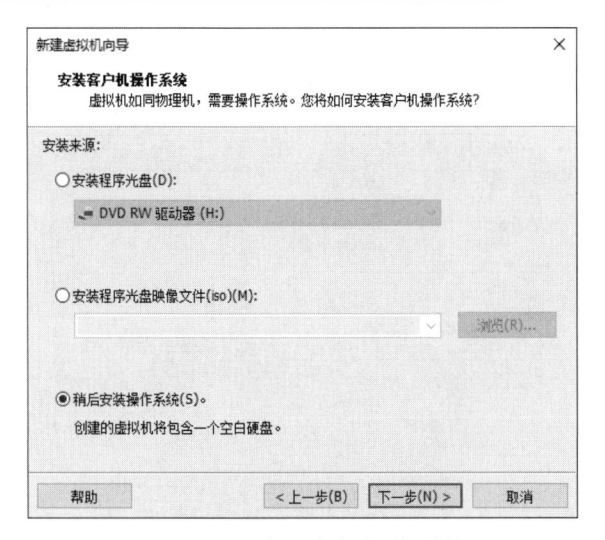

图 27.5　设置稍后安装操作系统

图 27.6　设置客户机操作系统

（10）在打开的"此虚拟机的内存"对话框中，将虚拟机的内存设置为 1024MB，如图 27.8 所示，单击"下一步"按钮。

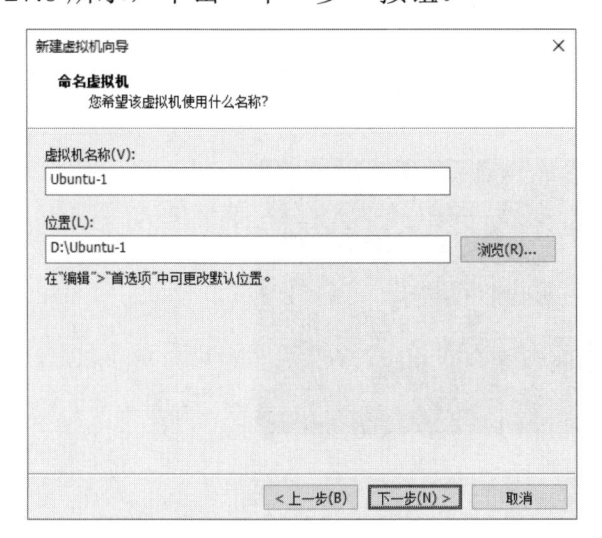

图 27.7　命名虚拟机

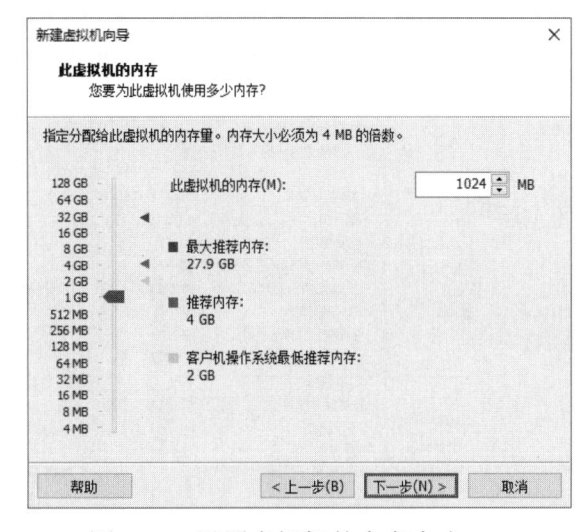

图 27.8　设置虚拟机的内存大小

（11）在打开的"网络类型"对话框中选中"使用网络地址转换（NAT）"单选按钮，单击"下一步"按钮。

（12）在打开的"选择 I/O 控制器类型"对话框中，单击"下一步"按钮。

（13）在打开的"选择磁盘"对话框中，单击"下一步"按钮。

（14）在打开的"指定磁盘容量"对话框中将最大磁盘大小设置为 20.0GB，并选中"将虚拟磁盘存储为单个文件"单选按钮，如图 27.9 所示，单击"下一步"按钮。

（15）在打开的"指定磁盘文件"对话框中，单击"下一步"按钮。

（16）在打开的"已准备好创建虚拟机"对话框中，可以看到虚拟机的摘要信息，如图 27.10 所示。单击"完成"按钮，至此，虚拟机创建完成。

143

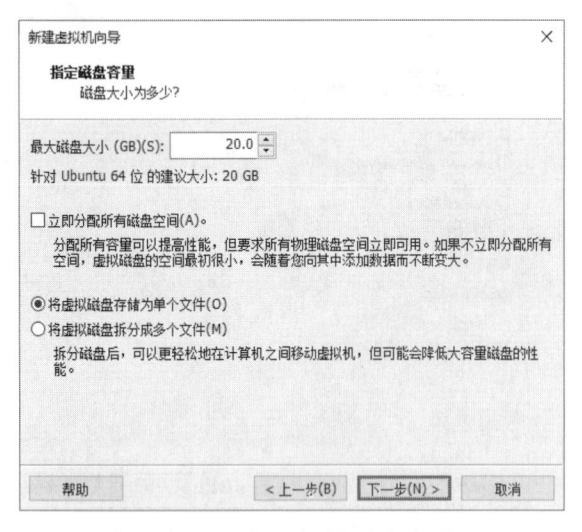

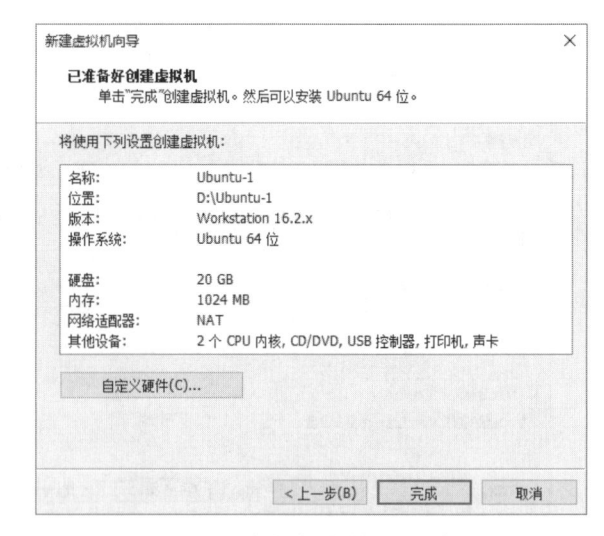

图 27.9　设置虚拟机的磁盘容量　　　　　图 27.10　虚拟机的摘要信息

（17）图 27.11 所示为虚拟机的硬件摘要信息和预览窗口。

图 27.11　虚拟机的硬件摘要信息和预览窗口

步骤 3：安装 Ubuntu 网络操作系统。

（1）选择虚拟机"Ubuntu-1"，在"Ubuntu-1"设备列表中双击光盘驱动器图标"CD/DVD（SATA）"，如图 27.12 所示。

（2）在打开的"虚拟机设置"对话框的"硬件"选项卡中，选择光盘驱动器"CD/DVD（SATA）"选项，选中"使用 ISO 映像文件"单选按钮，单击"浏览"按钮，如图 27.13 所示。

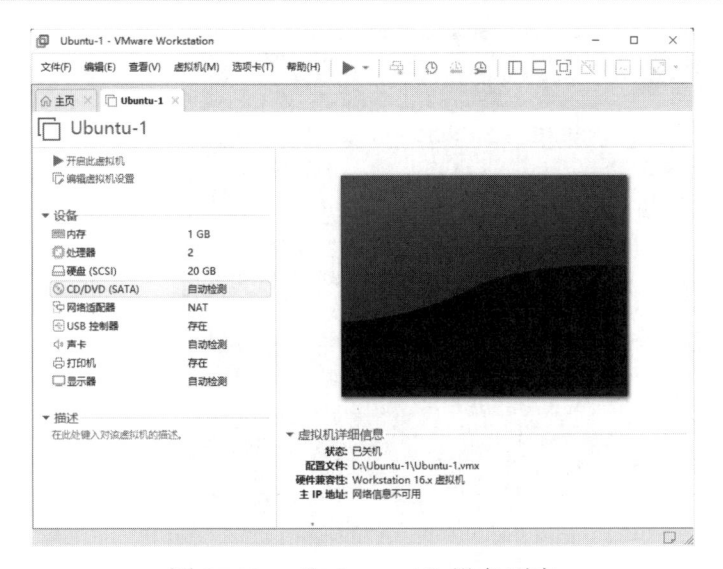

图 27.12　"Ubuntu-1"设备列表　　　　　图 27.13　设置光盘驱动器

（3）在打开的"浏览 ISO 映像"对话框中浏览并选择 Ubuntu 网络操作系统的安装映像文件，如图 27.14 所示，并单击"打开"按钮。

（4）返回"虚拟机设置"对话框，单击"确定"按钮完成设置。

（5）在虚拟机"Ubuntu-1"选项卡中，单击"开启此虚拟机"按钮来开启虚拟机，如图 27.15 所示。

图 27.14　选择安装映像文件　　　　　图 27.15　开启虚拟机

（6）加载后进入"GNU GRUB"界面，选择"*Try or Install Ubuntu Server"选项，如图 27.16 所示，并按"Enter"键即可开始安装。

（7）在引导和加载成功后出现欢迎界面，可以选择语言类型，这里选择"English"选项，如图 27.17 所示，并按"Enter"键。

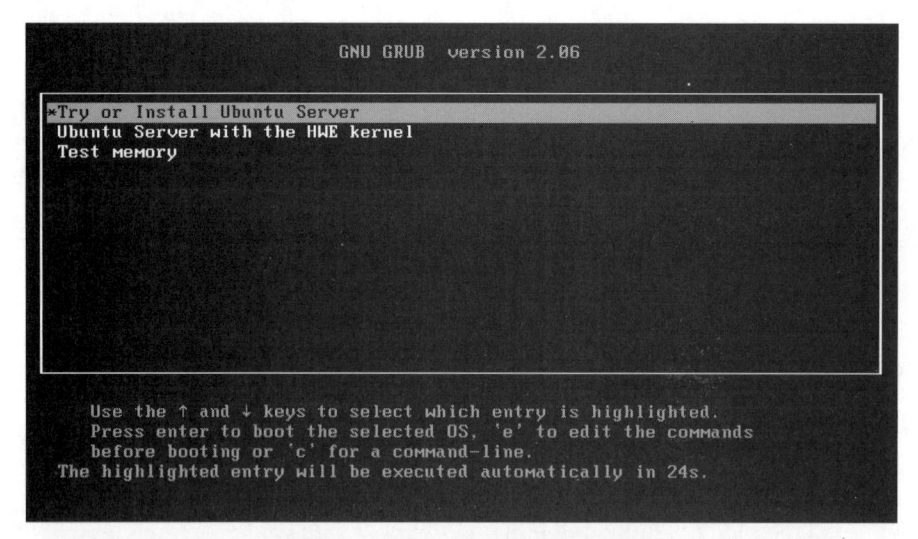

图 27.16　"GNU GRUB"界面

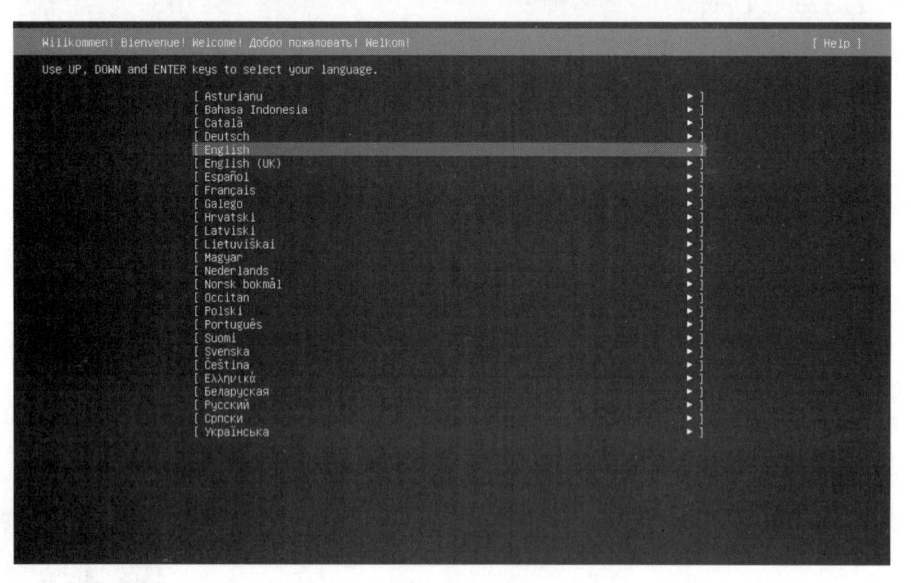

图 27.17　选择语言类型

（8）在打开的"Installer update available"界面中，使用默认配置，直接按"Enter"键。

（9）在打开的"Keyboard configuration"界面中，使用默认配置，直接按"Enter"键。

（10）在打开的"Choose type of install"界面中，使用默认配置，直接按"Enter"键。

（11）在打开的"Network connections"界面中，使用默认配置，直接按"Enter"键。

（12）在打开的"Configure proxy"界面中，使用默认配置，直接按"Enter"键。

（13）在打开的"Configure Ubuntu archive mirror"界面中，使用默认配置，直接按"Enter"键。

（14）在打开的"Mirror check still running"（镜像检测）界面中，选择"Continue"选项，如图 27.18 所示，并按"Enter"键。

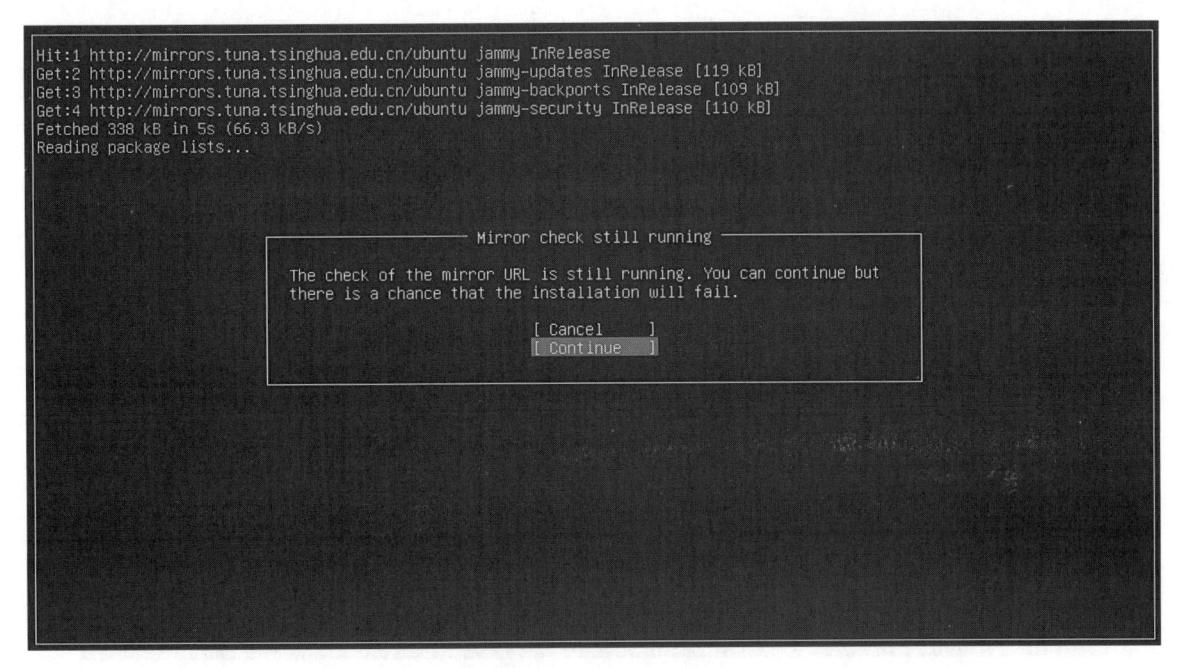

图 27.18　镜像检测界面

（15）在打开的"Guided storage configuration"界面中，使用默认配置，选择"Done"选项后直接按"Enter"键。

（16）在打开的"Storage configuration"界面中，使用默认配置，选择"Done"选项后直接按"Enter"键。

（17）在打开的"Confirm destructive action"（确认安装）界面中，选择"Continue"选项，如图 27.19 所示，并按"Enter"键。

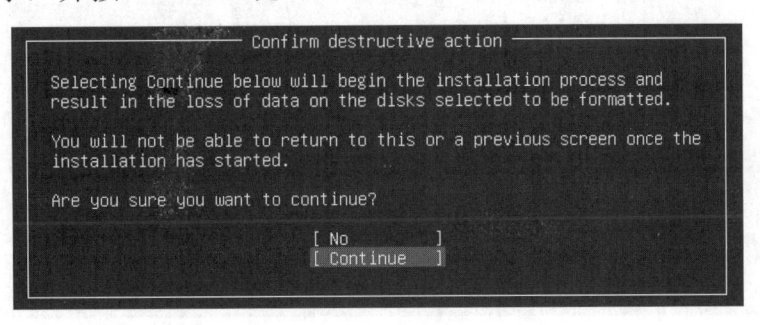

图 27.19　确认安装界面

（18）在打开的"Profile setup"界面中，输入用户登录系统的姓名、主机名、用户名和密码等，选择"Done"选项，如图 27.20 所示，并按"Enter"键，。

（19）在打开的"Upgrade to Ubuntu Pro"界面中，使用默认配置，选择"Continue"选项后直接按"Enter"键。

（20）在打开的"SSH Setup"界面中，先选择"Install OpenSSH server"选项，再选择"Done"选项，如图 27.21 所示，并按"Enter"键。

147

图 27.20 输入用户登录系统的姓名、主机名、用户名和密码等

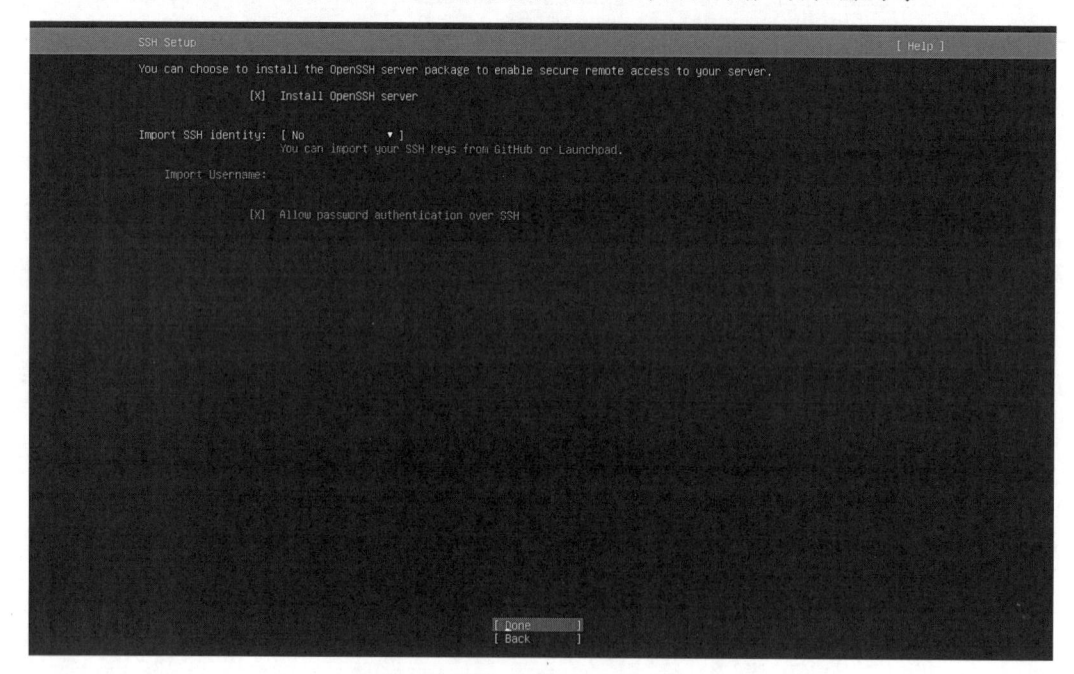

图 27.21 "SSH Setup"界面

（21）在打开的"Install complete！"界面中，表示安装已经完成，选择"Reboot Now"选项后直接按"Enter"键，安装完成界面如图 27.22 所示。

（22）在 Ubuntu 网络操作系统重新启动时，会提示"Failed unmounting/cdrom."信息，直接按"Enter"键即可。

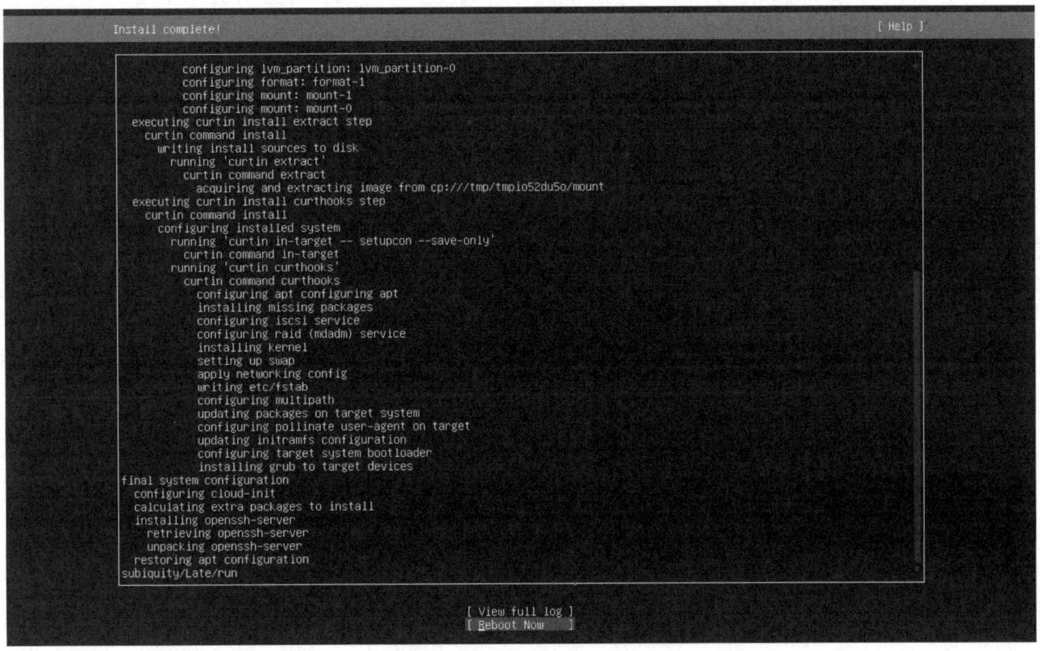

任务 27　安装与配置 Ubuntu 网络操作系统

图 27.22　安装完成界面

步骤 4：配置管理员 root 密码。

在安装 Ubuntu 网络操作系统的过程中，没有设置管理员 root 的用户名和密码。因此，在默认情况下，管理员 root 是无法登录系统的，需要单独配置管理员 root 的密码。打开终端窗口，可以配置管理员 root 密码，命令如下所示。

```
chris@ubuntu:~$ sudo passwd root
[sudo] password for chris:              //输入当前用户 chris 的密码
New password:                           //输入管理员 root 的密码
Retype new password:                    //再次输入管理员 root 的密码
passwd: password updated successfully
chris@ubuntu:~$ su root                 //以管理员 root 的身份登录
Password:                               //输入管理员 root 的密码
root@ubuntu:/home/chris#exit            //管理员 root 登录后的命令提示符为"#"
chris@ubuntu:~$                         //返回到用户 chris 的命令提示符下
```

步骤 5：查看 Ubuntu 网络操作系统相关信息。

（1）查看 Ubuntu 网络操作系统的内核版本信息，命令如下所示。

```
root@ubuntu:~# uname -a
Linux ubuntu 5.15.0-91-generic #101-Ubuntu SMP Tue Nov 14 13:30:08 UTC 2023 x86_64 x86_64 x86_64 GNU/Linux
```

或者：

```
root@ubuntu:~# cat /proc/version
Linux version 5.15.0-91-generic (buildd@lcy02-amd64-045) (gcc (Ubuntu 11.4.0-1ubuntu1~22.04) 11.4.0, GNU ld (GNU Binutils for Ubuntu) 2.38) #101-Ubuntu SMP Tue Nov 14 13:30:08 UTC 2023
```

（2）查看 Ubuntu 网络操作系统的版本信息，命令如下所示。

```
root@ubuntu:~# lsb_release -a
No LSB modules are available.
Distributor ID: Ubuntu                    //名称
Description:    Ubuntu 22.04.3 LTS        //版本号
Release:        22.04
Codename:       jammy
```

步骤 6： 查看 Ubuntu 网络操作系统防火墙的状态与版本。

（1）查看 Ubuntu 网络操作系统防火墙的状态，命令如下所示。

```
root@ubuntu:~# ufw status
Status: inactive
```

（2）查看 Ubuntu 网络操作系统防火墙的版本，命令如下所示。

```
root@ubuntu:~# ufw version
ufw 0.36.1
Copyright 2008-2021 Canonical Ltd.
```

任务 28　文件和目录管理

任务目标

1．掌握文件和目录的管理命令。
2．掌握 vim 编辑工具的三种模式。
3．使用文件和目录的管理命令进行创建、删除、复制和移动等操作。
4．使用 vim 编辑工具实现文件的操作。

任务描述

某公司的网络管理员小赵专心研究 Linux 操作系统的常用操作，找了很多资料后，决定从文件和目录管理开始学习。

任务要求

文件和目录管理是 Linux 操作系统基础命令中应用相对较多的命令，是广大初学者的首选学习内容。本任务的具体要求如下。

（1）在根目录下建立/test、/test/etc、/test/exer/task1、/test/exer/task2 目录，并使用"tree"命令查看/test 目录的结构。

（2）复制/etc 目录下所有以字母"h""i""j"开头的文件到/test/etc 目录下（包括子目录），将当前目录切换为/test/etc 目录，并以相对路径的方式查看/test/etc 目录下的内容。

（3）将当前目录切换为/test/exer/task1，在当前目录下建立空白文件 file1.txt 和 file2.txt，在当前目录下将 file2.txt 文件更名为"file4.txt"，以相对路径的方式将/test/etc/hosts 文件复制并粘贴为新文件/test/exer/task1/file3.txt，之后查看当前目录下的文件。

（4）以绝对路径的方式，直接删除/test/etc 目录下以"init"开头的所有文件或子目录，移动/test/etc 目录下以"is"开头的文件或子目录到/test/exer/task2 目录下。

（5）查看/test/etc 目录下以"host"开头的文件的文件类型。

（6）将当前目录切换为/test/exer/task2，使用相对路径的方式为 issue.net 文件建立硬链接文件 file5.txt，为 issue.net 文件建立软链接文件 file6.txt，将链接文件存放到/test/exer/task1 目录下，并查看两个目录中的文件列表。

（7）在用户家目录下，使用"echo"命令建立/var/info1 文件，文件内容如下所述。

```
Banana
Orange
Apple
```

（8）统计/etc/sysctl.conf 文件中的字节数、字数、行数，并将统计结果输出到/var/info2 文件中。

（9）使用命令查看/var/info1 文件前两行的内容，并将输出结果存放到/var/info3 文件中。

（10）使用命令查找/etc 目录下的文件名以"c"开头、以"conf"结尾且大于 5KB 的文件，将查询结果存放到/var/info4 文件中。

（11）使用命令输入/var/info1 文件的后两行，并将输出结果存放到/var/info5 文件中。

（12）使用命令输出/var/info1 文件中不包括"pp"字符串的行，并输出行号，将输出结果存放到/var/info6 文件中。

（13）在/root 目录下启动 vim，在"vim"命令的后面不加文件名。

（14）进入 vim 编辑模式，输入如下所示的测试文本。

```
Linux has the characteristics of open source, no copyright and
more users in the technology community.
Open source enables users to cut freely, with high flexibility, powerful function and
low cost.
In particular, the network protocol stack embedded in the system can realize the
function of router after proper configuration.
These characteristics make Linux an ideal platform for developing routing switching
devices.
```

（15）将以上文本保存为 Linux 文件，并退出 vim。

（16）重新启动 vim，打开 Linux 文件。

（17）显示文件行号。

（18）将鼠标光标移动到文件内容的第 4 行行首。

（19）在当前行下方插入新行，并输入内容"This is a very good system!"。

（20）将文本中的"Linux"用"Ubuntu"替换。

（21）将鼠标光标移动到文本的第 3 行行首，并复制第 3 行至第 4 行的内容。将鼠标光标移动到文本的最后一行行首，并将刚才复制的内容粘贴在最后一行的下方。

（22）保存文件后退出 vim。

（23）使用"tar"命令对用户家目录中的文件夹 test1 和文件 file1 打包，将打包文件命名为"1.tar",并存放到当前目录下。

(24)使用"tar"命令将 1.tar 文件恢复到/home 目录下。

(25)使用"tar"命令将 file2 文件追加到 tar 包的末尾。

任务实施

步骤 1:在根目录下建立/test、/test/etc、/test/exer/task1、/test/exer/task2 目录,安装并使用"tree"命令查看/test 目录的结构,命令如下所示。

```
root@ubuntu:~# mkdir -p /test/etc /test/exer/task1 /test/exer/task2
root@ubuntu:~# apt install -y tree              //安装"tree"命令
root@ubuntu:~# tree /test                        //以树状格式列出目录结构
/test
├── etc
└── exer
    ├── task1
    └── task2
4 directories, 0 files
```

步骤 2:复制/etc/目录下所有以字母"h""i""j"开头的文件到/test/etc 目录下(包括子目录),将当前目录切换为/test/etc 目录,并以相对路径的方式查看/test/etc 目录下的内容,命令如下所示。

```
root@ubuntu:~# cp -r /etc/[h-j]* /test/etc
root@ubuntu:~# cd /test/etc
root@ubuntu:/test/etc# ls
hdparm.conf  hostname  hosts.allow  init      initramfs-tools  iproute2  issue
host.conf    hosts     hosts.deny   init.d    inputrc                    iscsi     issue.net
```

步骤 3:将当前目录切换为/test/exer/task1 目录,在当前目录下建立空白文件 file1.txt 和 file2.txt。在当前目录下将 file2.txt 文件更名为"file4.txt",以相对路径的方式将/test/etc/hosts 文件复制并粘贴为新文件/test/exer/task1/file3.txt,之后查看当前目录下的文件,命令如下所示。

```
root@ubuntu:/test/etc# cd ../exer/task1
root@ubuntu:/test/exer/task1# touch file1.txt file2.txt
root@ubuntu:/test/exer/task1# mv file2.txt file4.txt
root@ubuntu:/test/exer/task1# cp ../../etc/hosts file3.txt
root@ubuntu:/test/exer/task1# ll
total 12
drwxr-xr-x 2 root root 4096    Jan 13 07:02  ./
drwxr-xr-x 4 root root 4096    Jan 13 05:54  ../
-rw-r--r-- 1 root root    0    Jan 13 07:00  file1.txt
```

```
-rw-r--r-- 1 root root     221 Jan 13 07:02  file3.txt
-rw-r--r-- 1 root root     0   Jan 13 07:00  file4.txt
```

步骤 4：以绝对路径的方式，直接删除/test/etc 目录下以"init"开头的所有文件或子目录，移动/test/etc 目录下以"is"开头的文件或子目录到/test/exer/task2 目录下，命令如下所示。

```
root@ubuntu:/test/exer/task1# rm -rf /test/etc/init*
root@ubuntu:/test/exer/task1# ls /test/etc
hdparm.conf       hostname      hosts.allow  inputrc   iscsi   issue.net
host.conf         hosts         hosts.deny   iproute2  issue
root@ubuntu:/test/exer/task1# mv /test/etc/is* /test/exer/task2
root@ubuntu:/test/exer/task1# ll /test/exer/task2
total 20
drwxr-xr-x 3 root root 4096    Jan 13 07:10  ./
drwxr-xr-x 4 root root 4096    Jan 13 05:54  ../
drwxr-xr-x 2 root root 4096    Jan 13 06:57  iscsi/
-rw-r--r-- 1 root root   26    Jan 13 06:57  issue
-rw-r--r-- 1 root root   19    Jan 13 06:57  issue.net
```

步骤 5：查看/test/etc 目录下以"host"开头的文件的文件类型，命令如下所示。

```
root@ubuntu:/test/exer/task1# file /test/etc/host*
/test/etc/host.conf:     ASCII text
/test/etc/hostname:      ASCII text
/test/etc/hosts:         ASCII text
/test/etc/hosts.allow:   ASCII text
/test/etc/hosts.deny:    ASCII text
```

步骤 6：将当前目录切换为/test/exer/task2 目录，使用相对路径的方式为 issue.net 文件建立硬链接文件 file5.txt，为 issue.net 文件建立软链接文件 file6.txt，将链接文件存放到/test/exer/task1 目录下，并查看两个目录中的文件列表，命令如下所示。

```
root@ubuntu:/test/exer/task1# cd ../task2
root@ubuntu:/test/exer/task2# pwd
/test/exer/task2
root@ubuntu:/test/exer/task2# ln issue ../task1/file5.txt
root@ubuntu:/test/exer/task2# ln -s issue.net ../task1/file6.txt
root@ubuntu:/test/exer/task2# ll -i
total 20
917511 drwxr-xr-x 3 root root 4096 Jan 13 07:10 ./
917509 drwxr-xr-x 4 root root 4096 Jan 13 05:54 ../
917584 drwxr-xr-x 2 root root 4096 Jan 13 06:57 iscsi/
917587 -rw-r--r-- 2 root root   26 Jan 13 06:57 issue
917588 -rw-r--r-- 1 root root   19 Jan 13 06:57 issue.net
root@ubuntu:/test/exer/task2# ll -i ../task1
total 16
```

```
917510 drwxr-xr-x 2 root root 4096 Jan 16 02:45 ./
917509 drwxr-xr-x 4 root root 4096 Jan 13 05:54 ../
917589 -rw-r--r-- 1 root root    0 Jan 13 07:00 file1.txt
917591 -rw-r--r-- 1 root root  221 Jan 13 07:02 file3.txt
917590 -rw-r--r-- 1 root root    0 Jan 13 07:00 file4.txt
917587 -rw-r--r-- 2 root root   26 Jan 13 06:57 file5.txt
917518 lrwxrwxrwx 1 root root    9 Jan 16 02:45 file6.txt -> issue.net
```

步骤7：在用户家目录下，使用"echo"命令建立/var/info1文件，文件内容如下所述。

```
Banana
Orange
Apple
```

命令如下所示。

```
root@ubuntu:~# pwd
/home/chris
root@ubuntu:~# echo Banana > info1
root@ubuntu:~# echo Orange >> info1
root@ubuntu:~# echo Apple >> info1
root@ubuntu:~# cat info1
Banana
Orange
Apple
```

步骤8：统计/etc/sysctl.conf文件中的字节数、字数、行数，并将统计结果存放到/var/info2文件中，命令如下所示。

```
root@ubuntu:~# pwd
/home/chris
root@ubuntu:~# wc /etc/sysctl.conf >info2
root@ubuntu:~# cat info2
 68  304 2351 /etc/sysctl.conf
```

步骤9：使用命令查看/var/info1文件前两行的内容，并将输出结果存放到/var/info3文件中，命令如下所示。

```
root@ubuntu:~# head -2 info1 >info3
root@ubuntu:~# cat info3
Banana
Orange
```

步骤10：使用命令查找/etc目录下的文件名以"c"开头、以"conf"结尾且大于5KB的文件，将查询结果存放到/var/info4文件中，命令如下所示。

```
root@ubuntu:~# find /etc -name "c*.conf" -size +5k >info4
root@ubuntu:~# cat info4
/etc/ca-certificates.conf
```

步骤11：使用命令输入/var/info1文件的后两行，并将输出结果存放到/var/info5文件中，命令如下所示。

```
root@ubuntu:~# tail -2 info1 >info5
root@ubuntu:~# cat info5
Orange
Apple
```

步骤 12：使用命令输出/var/info1 文件中不包括 "pp" 字符串的行，并输出行号，将输出结果存放到/var/info6 文件中，命令如下所示。

```
root@ubuntu:~# grep -n -v "pp" info1 >info6
root@ubuntu:~# cat info6
1:Banana
2:Orange
```

步骤 13：进入 Rocky Linux 8.6 操作系统界面，打开一个终端窗口，在命令行中输入"vim"（不加文件名）命令来启动 vim，并按"A"键进入编辑模式。

步骤 14：输入本任务要求（14）中的测试文本。

步骤 15：按"Esc"键返回到命令模式，输入":"命令进入末行模式，输入"w Linux"命令将程序保存为 Linux 文件，输入":q"命令退出 vim。

步骤 16：重新启动 vim，输入"vim Linux"命令打开 Linux 文件。

步骤 17：输入":set nu"命令来显示行号。

步骤 18：先按"3"键，再按"Enter"键，将鼠标光标移动到文件内容的第 4 行行首。

步骤 19：按"O"键在当前行下面添加新行，并输入内容"This is a very good system!"。

步骤 20：在编辑模式下按"Esc"键回到命令模式。输入":"命令进入末行模式，并输入"s/Linux/Ubuntu/g"命令将文本中的"Linux"替换成"Ubuntu"。

步骤 21：按"3"键并按"G"键，将鼠标光标移动到文本的第 3 行行首，输入"2yy"命令来复制第 3 行至第 4 行的内容。按"G"键，将鼠标光标移动到文本的最后一行行首，按"P"键将刚才复制的内容粘贴在最后一行的下方。

步骤 22：在末行模式下输入":wq"命令，保存文件后退出 vim。

步骤 23：在用户家目录中，使用"tar"命令将文件夹 test1 和文件 file1 打包，并将打包文件命名为"1.tar"，命令如下所示。

```
root@ubuntu:~# mkdir test1
root@ubuntu:~# touch file1
root@ubuntu:~# ls test1 file1
file1

test1:
```

```
root@ubuntu:~# tar -cvf 1.tar test1 file1
test1/
file1
root@ubuntu:~# ls 1.tar
1.tar
root@ubuntu:~# tar -tf 1.tar
test1/
file1
```

步骤 24：使用"tar"命令将 1.tar 文件恢复到/home 目录中，命令如下所示。

```
root@ubuntu:~# tar -xf 1.tar -C /test/
root@ubuntu:~# ls -d /test/test1 /test/file1
/test/file1  /test/test1
//从打包文件中恢复原文件时只需以-x选项代替-C选项即可
```

步骤 25：使用"tar"命令将 file2 文件追加到 tar 包的末尾，命令如下所示。

```
root@ubuntu:~# touch file2
root@ubuntu:~# tar -rf 1.tar file2
root@ubuntu:~# tar -tf 1.tar
test1/
file1
file2
//如果要将一个文件追加到tar包的末尾，则需要使用-r选项
```

任务 29 软件包管理

任务目标

1．了解 APT 的功能。
2．掌握"dpkg"命令的使用方法。
3．使用"apt"命令安装软件包。

任务描述

某公司的网络管理员小赵发现很多的软件包是 DEB 格式的，现在小赵需要对某些 DEB 格式的软件包进行安装，使用软件包管理工具可以实现快速安装。

任务要求

高级软件包管理工具（Advanced Packaging Tools，APT）是为了进一步降低软件安装难度和复杂度而设计的。高级软件包管理工具会自动计算软件包的依赖关系，并判断哪些软件应该安装，哪些软件不需要安装。使用高级软件包管理工具可以方便地进行软件的安装、查询、更新、卸载等，而且命令简洁、好记。

（1）使用"dpkg"命令查询 vim 软件包是否已安装。
（2）使用"dpkg"命令查询 vim 软件包的描述信息。
（3）配置 APT 软件源。
（4）使用"apt"命令安装 bind9 软件包。
（5）使用"apt"命令查询 bind9 软件包的安装情况。
（6）使用"apt"命令卸载 bind9 软件包。

任务实施

步骤 1：使用"dpkg"命令查询 vim 软件包是否已安装，命令如下所示。

```
root@ubuntu:~# dpkg -l vim
Desired=Unknown/Install/Remove/Purge/Hold
| Status=Not/Inst/Conf-files/Unpacked/halF-conf/Half-inst/trig-aWait/Trig-pend
|/ Err?=(none)/Reinst-required (Status,Err: uppercase=bad)
||/ Name           Version                    Architecture Description
+++-==============-==========================-============-===================
ii  vim            2:8.2.3995-1ubuntu2.15     amd64        Vi IMproved - enhanced vi editor
//ii 表示已安装成功
```

知识链接：

"dpkg"命令是一个安装、构建、删除和管理软件包的工具，命令格式如下所示。

dpkg [选项] 软件包名称

其中常用的选项如下。

- -i：表示安装软件包。
- -l：表示查询软件包的版本。
- -s：表示查询软件包的详细信息。

步骤2：使用"dpkg"命令查询vim软件包的描述信息，命令如下所示。

```
root@ubuntu:~# dpkg -s vim
Package: vim
Status: install ok installed
Priority: optional
Section: editors
Installed-Size: 3931
Maintainer: Ubuntu Developers <ubuntu-devel-discuss@lists.ubuntu.com>
Architecture: amd64
Version: 2:8.2.3995-1ubuntu2.15
Provides: editor
Depends: vim-common (= 2:8.2.3995-1ubuntu2.15), vim-runtime (= 2:8.2.3995-
1ubuntu2.15), libacl1 (>= 2.2.23), libc6 (>= 2.34), libgpm2 (>= 1.20.7), libpython3.10
(>= 3.10.0), libselinux1 (>= 3.1~), libsodium23 (>= 1.0.14), libtinfo6 (>= 6)
Suggests: ctags, vim-doc, vim-scripts
Description: Vi IMproved - enhanced vi editor
 Vim is an almost compatible version of the UNIX editor Vi.
 .
 Many new features have been added: multi level undo, syntax
 highlighting, command line history, on-line help, filename
 completion, block operations, folding, Unicode support, etc.
 .
 This package contains a version of vim compiled with a rather
 standard set of features.  This package does not provide a GUI
```

```
version of Vim.  See the other vim-* packages if you need more
(or less).
```

步骤 3：配置 APT 软件源。

由于 Ubuntu 网络操作系统的默认软件源地址服务器在欧洲，访问速度较慢，因此建议将默认软件源地址换成国内软件源地址。Ubuntu 网络操作系统的软件源配置文件为/etc/apt/sources.list。在国内的服务器中安装 Ubuntu 网络操作系统时，默认的 APT 软件源就是 Ubuntu 官方中国。当然，也可以改为其他软件源。例如，配置 API 软件源，将其改为阿里云的软件源，命令如下所示。

```
root@ubuntu:~# vim /etc/apt/sources.list
//以下内容是将文件中的内容替换后的内容，之后保存，并退出编辑器
deb https://mirrors.aliyun.com/ubuntu/ focal main restricted universe multiverse
deb-src https://mirrors.aliyun.com/ubuntu/ focal main restricted universe multiverse
deb https://mirrors.aliyun.com/ubuntu/ focal-security main restricted universe multiverse
deb-src https://mirrors.aliyun.com/ubuntu/ focal-security main restricted universe multiverse
deb https://mirrors.aliyun.com/ubuntu/ focal-updates main restricted universe multiverse
deb-src https://mirrors.aliyun.com/ubuntu/ focal-updates main restricted universe multiverse
# deb https://mirrors.aliyun.com/ubuntu/ focal-proposed main restricted universe multiverse
# deb-src https://mirrors.aliyun.com/ubuntu/ focal-proposed main restricted universe multiverse
deb https://mirrors.aliyun.com/ubuntu/ focal-backports main restricted universe multiverse
deb-src https://mirrors.aliyun.com/ubuntu/ focal-backports main restricted universe multiverse
root@ubuntu:~# apt update           //获取最新的软件包列表
root@ubuntu:~# apt upgrade          //更新当前系统中所有已安装的软件包
```

知识链接：

"apt"命令是 APT 提供的前端用户工具。与"apt-get"命令相比，"apt"命令进行了改进，增加了有用的选项和子命令，命令格式如下所示。

```
apt [选项] 子命令
```

其中常用的子命令如下。

- update：表示从软件仓库更新软件包索引。
- upgrade：表示升级软件包，但是不会删除软件包。

- install：表示安装软件包。
- remove：表示删除软件包。
- purge：表示彻底删除软件包。
- search：表示搜索软件包。
- show：表示显示软件包的信息。
- policy：表示显示软件包的安装状态和版本信息。

步骤4：使用"apt"命令安装bind9软件包，命令如下所示。

```
root@ubuntu:~# apt install -y bind9
```

步骤5：使用"apt"命令查询bind9软件包的安装情况，命令如下所示。

```
root@ubuntu:~# apt policy bind9
bind9:
  Installed: 1:9.18.18-0ubuntu0.22.04.1
  Candidate: 1:9.18.18-0ubuntu0.22.04.1
  Version table:
 *** 1:9.18.18-0ubuntu0.22.04.1 500
        500 http://cn.archive.ubuntu.com/ubuntu jammy-updates/main amd64 Packages
        100 /var/lib/dpkg/status
     1:9.18.12-0ubuntu0.22.04.3 500
        500 http://cn.archive.ubuntu.com/ubuntu jammy-security/main amd64 Packages
     1:9.18.1-1ubuntu1 500
        500 http://cn.archive.ubuntu.com/ubuntu jammy/main amd64 Packages
```

步骤6：使用"apt"命令卸载bind9软件包，命令如下所示。

```
root@ubuntu:~# apt remove -y bind9
```

知识链接：

执行"apt remove"命令可以卸载已安装的软件包，但会保留该软件包的配置文档。如果要同时删除该软件包的配置文件，则需要执行"apt purge"命令。

任务 30　配置常规网络与 SSH 服务

任务目标

1．掌握网络配置的相关配置文件和配置参数。
2．了解 SSH 服务的功能和原理。
3．熟练掌握 Linux 服务器网络的相关参数的配置方法。
4．掌握 SSH 配置和远程登录方法。

任务描述

某公司部署了若干台 Linux 服务器，网络管理员小赵按照公司的业务需求，对公司的 Linux 服务器进行配置并管理网络，实现 Linux 服务器与其他计算机的通信、远程连接安全管理信息中心内的服务器。

任务要求

如果 Linux 服务器想要与网络中的其他计算机进行通信、远程连接安全管理信息中心的服务器，首先要正确地进行网络配置。网络配置通常包括操作系统版本、主机名、IP 地址/子网掩码、默认网关、DNS 地址等，远程连接服务器同样需要配置。配置常规网络与 SSH 服务的网络拓扑结构，如图 30.1 所示。

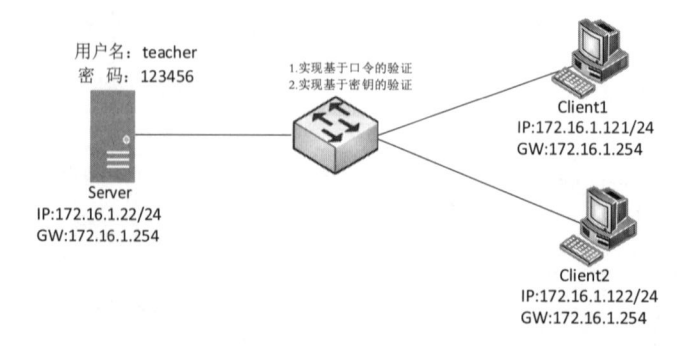

图 30.1　配置常规网络与 SSH 服务的网络拓扑结构

具体要求如下。

（1）两台计算机的配置如表 30.1 所示。

表 30.1　两台计算机的配置

项　　目	说　　明		
操作系统版本	Ubuntu	Windows 10	Ubuntu
主机	Server	Client1	Client2
IP 地址/子网掩码	172.16.1.22/24	172.16.1.121/24	172.16.1.122/24
默认网关	172.16.1.254		
DNS 地址	172.16.1.22、8.8.8.8		

（2）使用"ping"命令测试 Server 与 Client2 之间的连通性。

（3）在 Server 上，使用"netstat"命令查询处于监听状态的 TCP 连接。

（4）在 Client2 上，使用客户端软件连接远程服务器。

（5）使用基于密钥的验证方式进行 SSH 远程登录。

任务实施

步骤 1：配置 Server 的主机名，命令如下所示。

```
root@ubuntu:~# hostnamectl set-hostname server        //配置主机名
root@ubuntu:~# bash                                   //立即生效
root@server:~#
```

步骤 2：配置 Server 的 IP 地址为 172.16.1.22，子网掩码为 255.255.255.0，网关为 172.16.1.254，命令如下所示。

```
root@server:~# vim /etc/netplan/00-installer-config.yaml
network:
  ethernets:
    ens33:
        dhcp4: false                                  //关闭 DHCP
        addresses: [172.16.1.22/24]                   //配置静态 IP 地址/子网掩码
        gateway4: 172.16.1.254                        //配置网关
  version: 2
root@server:~# netplan apply                          //更新网络设置
```

步骤 3：配置 Server 的 DNS 地址为 172.16.1.22 和 8.8.8.8，命令如下所示。

```
root@server:~# vim /etc/resolv.conf
nameserver 172.16.1.22
nameserver 8.8.8.8
```

步骤 4：配置 Client1 与 Client2 的主机名、IP 地址和 DNS 地址等信息，参考 Server 的配置，此处不再赘述。

步骤 5：使用 "ping" 命令测试 Server 与 Client2 之间的连通性，命令如下所示。

```
root@server:~# ping 172.16.1.122      //当无任何选项时会一直测试，按"Ctrl+C"组合键停止
PING 172.16.1.122 (172.16.1.122) 56(84) bytes of data.
64 bytes from 172.16.1.122: icmp_seq=1 ttl=64 time=0.024 ms
64 bytes from 172.16.1.122: icmp_seq=2 ttl=64 time=0.033 ms
64 bytes from 172.16.1.122: icmp_seq=3 ttl=64 time=0.078 ms
64 bytes from 172.16.1.122: icmp_seq=4 ttl=64 time=0.043 ms
^C
--- 172.16.1.122 ping statistics ---
4 packets transmitted, 4 received, 0% packet loss, time 2999ms
rtt min/avg/max/mdev = 0.024/0.044/0.078/0.021 ms
```

步骤 6：在 Server 上，使用 "netstat" 命令查询处于监听状态的 TCP 连接，命令如下所示。

```
root@server:~# netstat -tl            //只显示处于监听状态的TCP连接
Active Internet connections (only servers)
Proto Recv-Q Send-Q Local Address       Foreign Address       State
tcp      0      0   0.0.0.0:ssh         0.0.0.0:*             LISTEN
tcp      0      0   localhost:domain    0.0.0.0:*             LISTEN
tcp6     0      0   [::]:ssh            [::]:*                LISTEN
root@server:~# netstat -tlan          //显示数字形式的地址，不转换为名称形式
Active Internet connections (servers and established)
Proto Recv-Q Send-Q Local Address       Foreign Address       State
tcp      0      0   0.0.0.0:22          0.0.0.0:*             LISTEN
tcp      0      0   127.0.0.53:53       0.0.0.0:*             LISTEN
tcp      0    372   172.16.1.22:22      172.16.1.122:52474    ESTABLISHED
tcp6     0      0   :::22               :::*                  LISTEN
```

步骤 7：使用客户端软件连接远程服务器，命令如下所示。

（1）在 Client2 上，安装 SecureCRT。

（2）在 Client2 上，打开 SecureCRT 主界面，如图 30.2 所示。

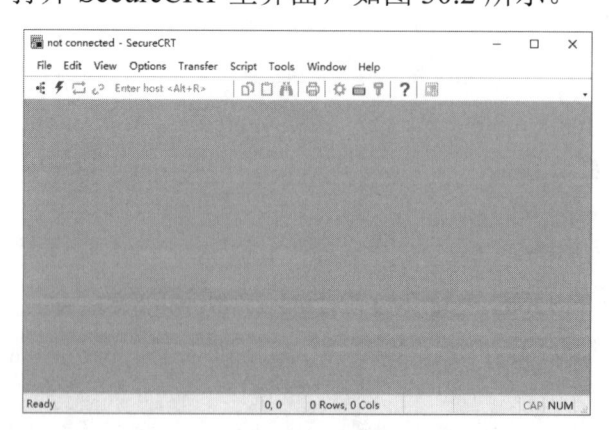

图 30.2　SecureCRT 主界面

（3）单击工具栏上的 "Quick Connect" 按钮，打开 "Quick Connect" 对话框，在

"Hostname"文本框中输入要连接的计算机的 IP 地址"172.16.1.22",在"Username"文本框中输入账号"chris",在"Authentication"选区中勾选"Password"复选框,其他选项保持默认设置即可,如图 30.3 所示,单击"Connect"按钮。

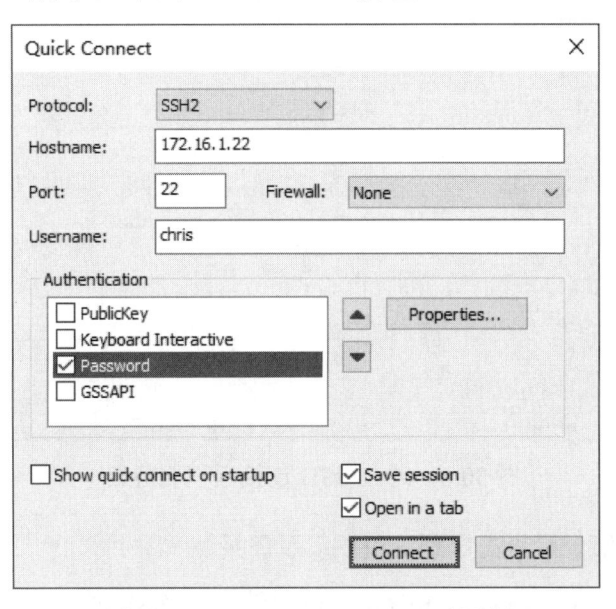

图 30.3 "Quick Connect"对话框

(4)在打开的"New Host Key"对话框中,如图 30.4 所示,单击"Accept & Save"按钮。

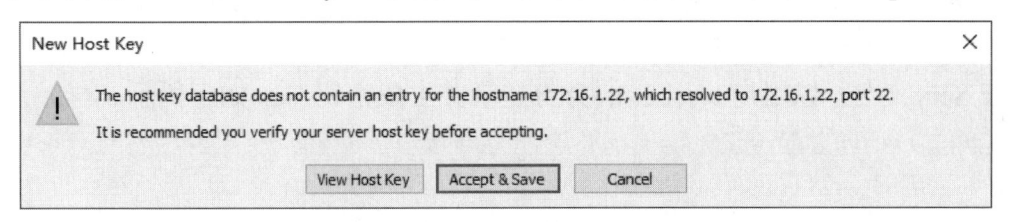

图 30.4 "New Host Key"对话框

(5)打开"Enter Secure Shell Password"对话框,在"Password"文本框中输入密码,如图 30.5 所示,单击"OK"按钮。

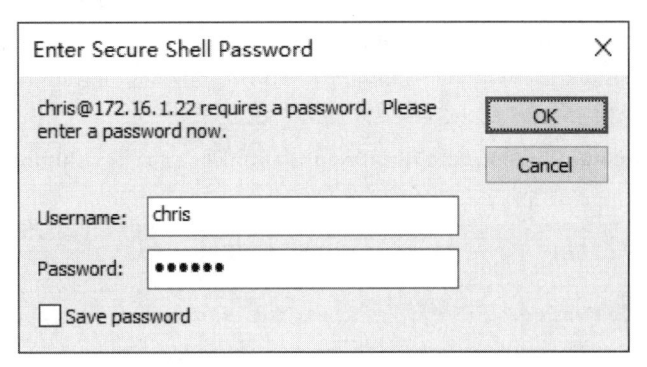

图 30.5 "Enter Secure Shell Password"对话框

(6)如果密码输入正确,则通过 SSH 连接远程服务器,如图 30.6 所示。如果密码输入错误，则再次打开"Enter Secure Shell Password"对话框，要求用户重新输入密码。

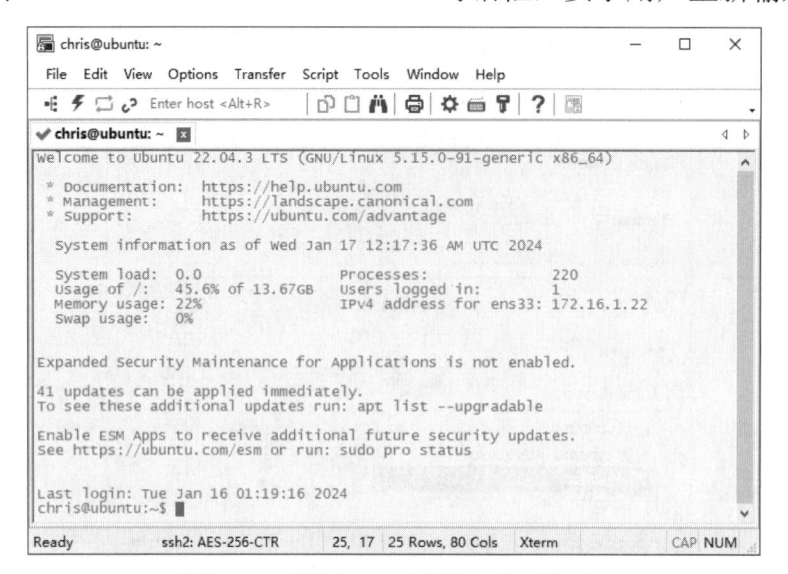

图 30.6　通过 SSH 连接远程服务器

步骤 8：基于密钥的验证方式实现 SSH 远程登录。

下面使用密钥验证方式，以用户 teacher 的身份登录 SSH 服务器，具体配置如下所示。

（1）在 Server 上，创建用户 teacher，并配置密码为"123456"，如下所示。

```
root@server:~# useradd -m teacher
root@server:~# passwd teacher
```

（2）在 Server 上，以用户 teacher 的身份登录并生成密钥对，如下所示。

```
$ ssh-keygen
Generating public/private rsa key pair.
Enter file in which to save the key (/home/teacher/.ssh/id_rsa)://按"Enter"键，或者配
置密钥的存储路径
Created directory '/home/teacher/.ssh'.
Enter passphrase (empty for no passphrase):     //直接按"Enter"键，或者配置密钥的密码
Enter same passphrase again:                    //再次按"Enter"键，或者配置密钥的密码
Your identification has been saved in /home/teacher/.ssh/id_rsa
Your public key has been saved in /home/teacher/.ssh/id_rsa.pub
The key fingerprint is:
SHA256:Ck5VxQPAEGcbmGvw389tBzAp5Zt72ePDh+uKE+eUxkE teacher@ubuntu
The key's randomart image is:
+---[RSA 3072]---+
|    o*=.o+.      |
|  . oo.+ + E     |
|   o .o o +      |
|   +. . = .      |
|   .o. S. * o    |
|   o ... + B     |
```

```
|   . . o O = .|
|      =.* B .|
|       .+.=+= |
+----[SHA256]-----+
```

(3) 在 Server 上,将生成的私钥文件传送至 Client2,命令如下所示。

```
$ scp ~/.ssh/id_rsa user1@172.16.1.122:/home/user1/.ssh/
The authenticity of host '172.16.1.122 (172.16.1.122)' can't be established.
ED25519 key fingerprint is SHA256:qu369pT+A8xXrYEovNYZz1+OFdGhYnJSdNDE+wGeayA.
This key is not known by any other names
Are you sure you want to continue connecting (yes/no/[fingerprint])? yes
Warning: Permanently added '172.16.1.122' (ED25519) to the list of known hosts.
user1@172.16.1.122's password:          //此处输入 Client2 的用户 user1 的密码
id_rsa                              100% 2602     2.8Mbit/s   00:00
```

(4) 在 Server 上,将生成的公钥文件保存在 authorized_keys 文件中,命令如下所示。

```
$ cd ~/.ssh
$ cat id_rsa.pub>authorized_keys
```

(5) 在 Client2 上,将 Server 传输过去的 id_rsa 私钥文件保存在 authorized_keys 文件中,命令如下所示。

```
user1@ubuntu:~$ cd ~/.ssh
user1@ubuntu:~/.ssh$ cat id_rsa>authorized_keys
```

(6) 在 Server 上,取消第 57 行代码的注释,并将 "PasswordAuthentication yes" 改为 "PasswordAuthentication no",使其不允许公钥验证,拒绝传送的口令验证方式。保存后退出并重启 sshd 服务程序,命令如下所示。

```
root@server:~# vim /etc/ssh/sshd_config
……                //此处省略部分内容
   56 # To disable tunneled clear text passwords, change to no here!
   57 PasswordAuthentication no
   58 #PermitEmptyPasswords no
……                //此处省略部分内容
root@server:~# systemctl restart sshd
```

任务验收

(1) 查看系统主机名,命令如下所示。

```
root@server:~# cat /etc/hostname            //查看系统主机名
server.phei.com.cn
```

(2) 查看网卡配置信息,命令如下所示。

```
root@server:~# ip addr show ens33           //查看 IP 地址信息
```

```
2: ens33: <BROADCAST,MULTICAST,UP,LOWER_UP> mtu 1500 qdisc fq_codel state UP group
default qlen 1000
    link/ether 00:0c:29:3f:f6:df brd ff:ff:ff:ff:ff:ff
    altname enp2s1
    inet 172.16.1.22/24 metric 100 brd 172.16.1.255 scope global dynamic ens33
       valid_lft 1776sec preferred_lft 1776sec
    inet6 fe80::20c:29ff:fe3f:f6df/64 scope link
       valid_lft forever preferred_lft forever
```

（3）在 Client1 上，测试 SSH 能否以基于密钥的方式登录服务器。

在 Client1 上，尝试以用户 teacher 的身份远程登录服务器，此时无须输入密码也可以成功登录。同时使用"ip addr"命令可以看到 Server 网卡的 IP 地址为 172.16.1.22，说明已成功登录到了远程服务器上。

```
user1@ubuntu:~$ ssh teacher@172.16.1.22
Welcome to Ubuntu 22.04.3 LTS (GNU/Linux 5.15.0-91-generic x86_64)
$ip addr show ens33
2: ens33: <BROADCAST,MULTICAST,UP,LOWER_UP> mtu 1500 qdisc fq_codel state UP group
default qlen 1000
    link/ether 00:0c:29:3f:f6:df brd ff:ff:ff:ff:ff:ff
    altname enp2s1
    inet 172.16.1.22/24 metric 100 brd 172.16.1.255 scope global dynamic ens33
       valid_lft 1254sec preferred_lft 1254sec
    inet6 fe80::20c:29ff:fe3f:f6df/64 scope link
       valid_lft forever preferred_lft forever
```

在 Server 上，查看 Client1 的公钥是否传送成功，如下所示。

```
root@server:~# cat /home/teacher/.ssh/authorized_keys
ssh-rsa
AAAAB3NzaC1yc2EAAAADAQABAAABgQC/DmI8QWl65xXt2lG0M+V6T5fhjNVAd37dLIswyJULSfCcVz3XKGk/xihb
CS3+7pja9tFk/2XPhNuY4yKSV+bcJdPniavbrYyZMxfUWilEtRJ9a5+lJj0l0APDezjkTpb6K6I8cpQOXn6vnZLR
pyxPDWVTQHbKZ3OvqzoglM7gp10PHi9VGDaiE7EVPvnwgCePheGFhyuwLJMC25jrruM+rImLUO/CvJF49DuMp6+y
BLAmi92Hcf0dmSg1AmSzNGHvyVvJeDEr3cS3VJVSzT0s47rNUGvjF6fRRlNTVHMBB+hR1ehz8f7nGNfFpSJGcLa0
MKQ7izfRDUnz2RyQ0+T2DE8WTypJdNRGHJRcDtXRdd5LDWkKmPFpWwr4SXdbn7Bb93lmfib+tShKSzjX5NHreYBT
StjmrFnCbQg2WuLwp8zT8S8H+zuX1DZKvCImHxu5pVLzIyFO1jPM1NhM12uEvnaGuKm52OPem4f7cGJzdxdFT/x2
sKzfCb4nKhjGw3E= teacher@ubuntu
```

任务 31 磁盘管理

任务目标

1. 掌握磁盘的管理命令。
2. 使用"fdisk""mkfs"等磁盘管理命令对磁盘进行分区与格式化。
3. 正确使用创建文件系统命令和分区挂载命令。

任务描述

某公司购置了 Linux 服务器,现在网络管理员小赵需要对 Linux 操作系统中的磁盘进行分区,并创建不同类型的磁盘格式。在 Linux 操作系统中,需要将不同类型的文件系统挂载在不同的分区下,并使用命令查看磁盘的使用情况,验证磁盘管理的正确性。

任务要求

分区从实质上来说就是对磁盘的一种格式化,磁盘只有分区和格式化后才能使用,在 Linux 操作系统中可以使用"fdisk"和"mkfs"命令实现,具体要求如下。

(1) 添加一块大小为 20GB 的磁盘。
(2) 将 MBR 分区方式转换成 GPT 分区方式。
(3) 使用"fdisk"磁盘管理命令来创建 2 个主分区和 2 个逻辑分区,每个主分区大小为 5GB;逻辑分区大小分别为 8GB 和 2GB。
(4) 将创建好的分区格式化,格式化的文件系统为 ext4。
(5) 对格式化后的磁盘分区设置自动挂载。
(6) 验证磁盘分区和自动挂载。

任务实施

步骤 1：为虚拟机添加磁盘。

（1）在进行磁盘管理前需要添加一块磁盘。在虚拟机中添加磁盘非常容易，在虚拟机界面中单击"编辑虚拟机设置"按钮，打开"虚拟机设置"对话框，如图 31.1 所示。

图 31.1 "虚拟机设置"对话框

（2）单击"添加"按钮，打开"添加硬件向导"的"硬件类型"对话框，在"硬件类型"选区中选择"硬盘"选项，如图 31.2 所示。

图 31.2 "添加硬件向导"的"硬件类型"对话框

（3）单击"下一步"按钮，在打开的"添加硬件向导"对话框中设置磁盘类型为"SCSI"，并选中"创建新虚拟磁盘"单选按钮。打开"添加硬件向导"的"指定磁盘容量"对话框，指定磁盘大小为 20GB，并选中"将虚拟磁盘存储为单个文件"单选按钮，如图 31.3 所示，单击"下一步"按钮。在打开的"添加硬件向导"的"指定磁盘文件"对话框中设置磁盘文件的存储位置，如图 31.4 所示。单击"完成"按钮，磁盘添加完成效果如图 31.5 所示。

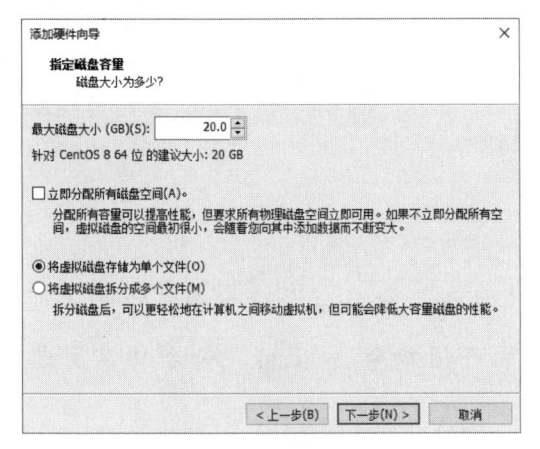

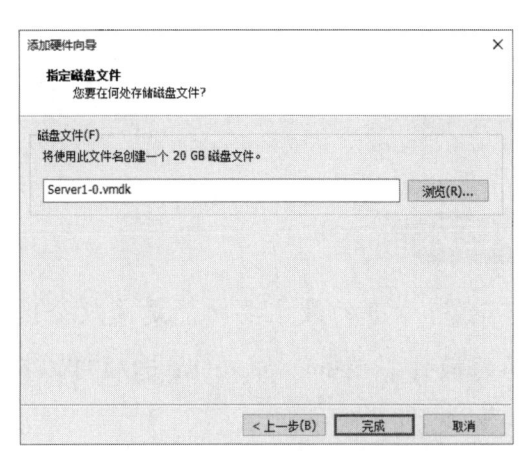

图 31.3　设置磁盘大小

图 31.4　设置磁盘存储位置

图 31.5　磁盘添加完成效果

步骤 2：使用"fdisk"命令管理磁盘分区方式。

（1）使用"fdisk"命令查看磁盘信息，如下所示。

```
root@ubuntu:~# fdisk -l
Disk /dev/loop0: 111.95 MiB, 117387264 bytes, 229272 sectors
Units: sectors of 1 * 512 = 512 bytes
Sector size (logical/physical): 512 bytes / 512 bytes
```

```
......                                          //此处省略部分内容
Disk /dev/sdb: 20 GiB, 21474836480 bytes, 41943040 sectors
Disk model: VMware Virtual S
Units: sectors of 1 * 512 = 512 bytes
Sector size (logical/physical): 512 bytes / 512 bytes
I/O size (minimum/optimal): 512 bytes / 512 bytes
......                                          //此处省略部分内容
//可以看出/dev/sdb是新添加的硬盘，是没有经过分区和格式化的
```

知识链接：

"fdisk"命令的最主要功能是修改分区表。在"fdisk"命令后面输入相应的选项可以进行不同的操作。例如，输入 m 选项可以列出所有的可用命令。"fdisk"命令的常用选项及其功能如表31.1所示。

表31.1 "fdisk"命令的常用选项及其功能

选 项	功 能	选 项	功 能
a	调整硬盘启动分区	q	不保存更改，退出 fdisk 操作菜单
d	删除硬盘分区	t	更改分区类型
l	列出所有支持的分区类型	u	切换所显示的分区大小的单位
m	列出所有的可用命令	w	先把修改写入硬盘分区表，再退出
n	创建新分区	x	列出高级选项
p	列出硬盘分区表		

（2）将磁盘 MBR 分区方式转换成 GPT 分区方式，如下所示。

```
root@ubuntu:~# parted /dev/sdb
GNU Parted 3.4
Using /dev/sdb
Welcome to GNU Parted! Type 'help' to view a list of commands.
(parted) mklabel  gpt                 //设置磁盘为GPT分区方式
Warning: The existing disk label on /dev/sdb will be destroyed and all data on this disk will be lost. Do you want to continue?
Yes/No? yes                           //确认修改
(parted) quit                         //退出
Information: You may need to update /etc/fstab.
root@ubuntu:~# fdisk /dev/sdb         //查看磁盘格式

Command (m for help): p
Disk /dev/sdb: 20 GiB, 21474836480 bytes, 41943040 sectors
Disk model: VMware Virtual S
Units: sectors of 1 * 512 = 512 bytes
Sector size (logical/physical): 512 bytes / 512 bytes
```

```
I/O size (minimum/optimal): 512 bytes / 512 bytes
Disklabel type: gpt                            //磁盘已转换成 GPT 分区方式
Disk identifier: 7381377D-38AE-4E0A-A1E6-48F910BE6B63
```

（3）创建主分区。

使用如下所示命令，打开 fdisk 操作菜单。

```
root@ubuntu:~# fdisk /dev/sdb
```

输入 p 选项，列出当前硬盘分区表。从命令执行结果可以看到，/dev/sdb 硬盘并无任何分区，如下所示。

```
Command (m for help): p
Disk /dev/sdb: 20 GiB, 21474836480 bytes, 41943040 sectors
Disk model: VMware Virtual S
Units: sectors of 1 * 512 = 512 bytes
Sector size (logical/physical): 512 bytes / 512 bytes
I/O size (minimum/optimal): 512 bytes / 512 bytes
Disklabel type: gpt
Disk identifier: 7381377D-38AE-4E0A-A1E6-48F910BE6B63
```

先输入 n 选项，再输入 p 选项，并且创建编号为 1 和 2 的主分区，2 个主分区大小均为 5GB，如下所示。

```
Command (m for help): n
Partition type                                         //分区类型
   p   primary (0 primary, 0 extended, 4 free)         //主分区
   e   extended (container for logical partitions)     //扩展分区
Select (default p): p
Partition number (1-4, default 1): 1
First sector (2048-41943039, default 2048):
Last sector, +/-sectors or +/-size{K,M,G,T,P} (2048-41943039, default 41943039): +5G

Created a new partition 1 of type 'Linux' and of size 5 GiB.
//创建了一个新分区 1，类型为"Linux"，大小为 5 GB


Command (m for help): n
Partition type
   p   primary (1 primary, 0 extended, 3 free)
   e   extended (container for logical partitions)
Select (default p): p
Partition number (2-4, default 2): 2
First sector (10487808-41943039, default 10487808):
Last sector, +/-sectors or +/-size{K,M,G,T,P} (10487808-41943039, default 41943039): +5G

Created a new partition 2 of type 'Linux' and of size 5 GiB.
```

（4）创建扩展分区。

创建编号为 3 的扩展分区，将剩余空间全部分给扩展分区，起始柱面和结束柱面全部保持默认设置，按"Enter"键即可，如下所示。

```
Command (m for help): n
Partition type
   p   primary (2 primary, 0 extended, 2 free)
   e   extended (container for logical partitions)
Select (default p): e
Partition number (3,4, default 3): 3
First sector (20973568-41943039, default 20973568):
Last sector, +/-sectors or +/-size{K,M,G,T,P} (20973568-41943039, default 41943039):

Created a new partition 3 of type 'Extended' and of size 10 GiB.
//分区 3 已被设置为 Extended 类型，大小为 10 GB
```

（5）创建逻辑分区。

在扩展分区上创建逻辑分区，其中一个空间大小为 8GB，剩下空间全部分给另一个逻辑分区，逻辑分区无须指定编号，如下所示。

```
Command (m for help): n
All space for primary partitions is in use.
Adding logical partition 5
First sector (20975616-41943039, default 20975616):
Last sector, +/-sectors or +/-size{K,M,G,T,P} (20975616-41943039, default 41943039): +8G

Created a new partition 5 of type 'Linux' and of size 8 GiB.
//分区 5 已被设置为 Linux 类型，大小为 8 GB

Command (m for help): n
All space for primary partitions is in use.
Adding logical partition 6
First sector (37754880-41943039, default 37754880):
Last sector, +/-sectors or +/-size{K,M,G,T,P} (37754880-41943039, default 41943039):

Created a new partition 6 of type 'Linux' and of size 2 GiB.
//分区 6 已被设置为 Linux 类型，大小为 2 GB
```

（6）查看分区结果。

在完成全部分区后，可以输入 p 选项列出硬盘分区表，还可以先输入 w 选项将新添加的分区表写入硬盘分区表，再退出，否则新的分区表不能生效，如下所示。

```
Command (m for help): p
Disk /dev/sdb: 20 GiB, 21474836480 bytes, 41943040 sectors
Disk model: VMware Virtual S
Units: sectors of 1 * 512 = 512 bytes
```

```
Sector size (logical/physical): 512 bytes / 512 bytes
I/O size (minimum/optimal): 512 bytes / 512 bytes
Disklabel type: dos
Disk identifier: 0xa9f12cc2

Device     Boot    Start      End       Sectors    Size  ID  Type
/dev/sdb1          2048       10487807  10485760   5G    83  Linux
/dev/sdb2          10487808   20973567  10485760   5G    83  Linux
/dev/sdb3          20973568   41943039  20969472   10G   5   Extended
/dev/sdb5          20975616   37752831  16777216   8G    83  Linux
/dev/sdb6          37754880   41943039  4188160    2G    83  Linux

Command (m for help): w
The partition table has been altered.
Calling ioctl() to re-read partition table.
Syncing disks.
//正在同步磁盘
```

步骤 3：使用"mkfs"命令创建文件系统。

使用"mkfs.xfs /dev/sdb1"命令将主分区/dev/sdb1 格式换成 XFS 分区（sdb2、sdb5 和 sdb6 操作省略），如下所示。

```
root@ubuntu:~# mkfs.xfs /dev/sdb1
meta-data=/dev/sdb1              isize=512    agcount=4, agsize=327680 blks
         =                       sectsz=512   attr=2, projid32bit=1
         =                       crc=1        finobt=0, sparse=0
data     =                       bsize=4096   blocks=1310720, imaxpct=25
         =                       sunit=0      swidth=0 blks
naming   =version 2              bsize=4096   ascii-ci=0 ftype=1
log      =internal log           bsize=4096   blocks=2560, version=2
         =                       sectsz=512   sunit=0 blks, lazy-count=1
realtime =none                   extsz=4096   blocks=0, rtextents=0
```

知识链接：

"mkfs"命令用于创建文件系统，语法格式如下所示。

mkfs [选项] 分区设备名称

"mkfs"命令的常用选项及其功能如表 31.2 所示。

表 31.2　"mkfs"命令的常用选项及其功能

选　项	功　能
-t	指定要创建的文件系统类型
-c	创建文件系统前先检查坏块
-v	显示创建文件系统的详细信息

步骤4：分区手动挂载。

（1）将/dev/sdb1 挂载到/data1、/dev/sdb2 挂载到/data2、/dev/sdb5 挂载到/data3、/dev/sdb6 挂载到/data4，具体操作如下所示。

```
root@ubuntu:~# mkdir /data1 /data2 /data3 /data4
root@ubuntu:~# mount /dev/sdb1 /data1
root@ubuntu:~# mount /dev/sdb2 /data2
root@ubuntu:~# mount /dev/sdb5 /data3
root@ubuntu:~# mount /dev/sdb6 /data4
```

（2）在挂载成功后，可以通过"mount|grep sdb"命令查看挂载信息，如下所示。

```
root@ubuntu:~# mount|grep sdb
/dev/sdb1 on /data1 type xfs
(rw,relatime,attr2,inode64,logbufs=8,logbsize=32k,noquota)
/dev/sdb2 on /data2 type xfs
(rw,relatime,attr2,inode64,logbufs=8,logbsize=32k,noquota)
/dev/sdb5 on /data3 type xfs
(rw,relatime,attr2,inode64,logbufs=8,logbsize=32k,noquota)
/dev/sdb6 on /data4 type xfs
(rw,relatime,attr2,inode64,logbufs=8,logbsize=32k,noquota)
```

💡 小提示

> 当设备挂载到指定的挂载点目录时，挂载点目录下的原有文件会被暂时隐藏，无法访问。此时在挂载点目录下显示的是设备上的文件。当设备卸载后，挂载点目录下的文件将恢复显示。

步骤5：对格式化后的磁盘分区设置自动挂载。

（1）在系统每次运行时，实现分区自动挂载，可以在/etc/fstab 文件中将/dev/sdb1 分区以 defaults 方式挂载到/data1 挂载点，添加内容如下所示。

```
root@ubuntu:~# cat /etc/fstab
……                                   //此处省略部分内容
/swap.img       none       swap       sw       0   0
/dev/sdb1       /data1     xfs        defaults 0   0
```

（2）重启计算机，可以通过"mount|grep sdb1"命令查看挂载信息，如下所示。

```
root@ubuntu:~# mount|grep sdb1
/dev/sdb1 on /data1 type xfs
(rw,relatime,attr2,inode64,logbufs=8,logbsize=32k,noquota)
```

任务 32　创建与管理软 RAID

任务目标

1. 了解 RAID 的作用。
2. 掌握 RAID 的工作原理。
3. 能使用"mdadm"命令完成软 RAID 的配置。

任务描述

某公司的网络管理员小赵最近在访问服务器时，感觉访问速度慢，经过排查发现服务器的磁盘空间即将用完，小赵决定添置大容量磁盘为服务器提供网络存储、文件共享、数据库等网络服务功能，以满足日常办公需求。为了解决速度慢、空间不够等问题，小赵决定购买磁盘后使用 RAID（Redundant Arrays of Independent Disks，独立磁盘冗余阵列）进行管理。

任务要求

RAID 管理是基于卷的管理。卷是由一个或多个磁盘上的可用空间组成的存储单元，可以将其格式化为一种文件系统并分配驱动器号。动态磁盘具有提供容错、提高磁盘利用率和访问效率的功能。本任务的具体要求如下所示。

（1）添加四块磁盘，每块磁盘大小为 5GB。
（2）使用"mdadm"命令为前三块磁盘分别创建软 RAID 5，并将三块磁盘的软 RAID 5 分别命名为"/dev/md0""/dev/md1""/dev/md2"。
（3）挂载创建好的软 RAID 5。
（4）如果有一块磁盘已经损坏，则更换第四块磁盘作为新的软 RAID 5。

任务实施

步骤 1： 创建与挂载软 RAID 5。

（1）在虚拟机中添加四块磁盘，每块磁盘大小为5GB，具体步骤参考任务31。

（2）使用"fdisk"命令查看添加的磁盘情况，如下所示。

```
root@ubuntu:~# fdisk -l|grep /dev/sd
Disk /dev/sda: 30 GiB, 32212254720 bytes, 62914560 sectors
/dev/sda1        2048     4095     2048   1M BIOS boot
/dev/sda2        4096  4198399  4194304   2G Linux filesystem
/dev/sda3     4198400 62912511 58714112  28G Linux filesystem
Disk /dev/sdb: 5 GiB, 5368709120 bytes, 10485760 sectors
Disk /dev/sdc: 5 GiB, 5368709120 bytes, 10485760 sectors
Disk /dev/sdd: 5 GiB, 5368709120 bytes, 10485760 sectors
Disk /dev/sde: 5 GiB, 5368709120 bytes, 10485760 sectors
```

（3）使用"mdadm"命令将前三块磁盘创建为软RAID 5，设置软RAID 5的名称为"/dev/mdX"，其中"X"为设备编号，编号从"0"开始，如下所示。

```
root@ubuntu:~# mdadm -Cv /dev/md0 -a yes -n 3 -l 5 /dev/sdb /dev/sdc /dev/sdd
mdadm: layout defaults to left-symmetric
mdadm: layout defaults to left-symmetric
mdadm: chunk size defaults to 512K
mdadm: size set to 5237760K
mdadm: Defaulting to version 1.2 metadata
mdadm: array /dev/md0 started
```

（4）为/dev/md0创建类型为XFS的文件系统，如下所示。

```
root@ubuntu:~# mkfs -t xfs /dev/md0
log stripe unit (524288 bytes) is too large (maximum is 256KiB)
log stripe unit adjusted to 32KiB
meta-data=/dev/md0              isize=512    agcount=16, agsize=163712 blks
         =                       sectsz=512   attr=2, projid32bit=1
         =                       crc=1        finobt=1, sparse=1, rmapbt=0
         =                       reflink=1    bigtime=0 inobtcount=0
data     =                       bsize=4096   blocks=2618880, imaxpct=25
         =                       sunit=128    swidth=256 blks
naming   =version 2              bsize=4096   ascii-ci=0, ftype=1
log      =internal log           bsize=4096   blocks=2560, version=2
         =                       sectsz=512   sunit=8 blks, lazy-count=1
realtime =none                   extsz=4096   blocks=0, rtextents=0
```

（5）查看/dev/md0的具体情况，如下所示。

```
root@ubuntu:~# mdadm -D /dev/md0
/dev/md0:
......                       //此处省略部分内容
    Number   Major   Minor   RaidDevice  State
       0       8       16        0       active sync   /dev/sdb
       1       8       32        1       active sync   /dev/sdc
```

```
     3       8       64        2          active sync   /dev/sdd
```

> **知识链接：**

"mdadm"命令用于管理 Linux 操作系统中的软 RAID，其基本语法格式如下所示。

mdadm [模式] RAID 设备 [选项] 成员设备名称

当前，生产环境中用到的服务器一般都会配备软 RAID，如果没有 RAID 阵列卡，则必须使用"mdadm"命令在 Linux 操作系统中创建和管理软 RAID。"mdadm"命令的常用选项及其功能如表 32.1 所示。

表 32-1 "mdadm"命令的常用选项及其功能

选 项	功 能
-a	检测设备名称
-n	指定设备数量
-l	指定软 RAID 等级
-C	创建软 RAID
-v	显示过程
-f	模拟设备损坏
-r	移除设备
-Q	查看摘要信息
-D	查看详细信息
-S	停止软 RAID

（6）挂载软 RAID 5。将/dev/md0 挂载到指定的/media/md0 目录，挂载成功后可以看到可用空间为 9.9 GB，如下所示。

```
root@ubuntu:~# mkdir /media/md0
root@ubuntu:~# mount /dev/md0 /media/md0
root@ubuntu:~# df -h|grep /dev/md0
/dev/md0              10G    105M   9.9G   2%  /media/md0
```

步骤 2：修复软 RAID 5。

在生产环境中部署软 RAID 5，是为了提高磁盘的读写速度及数据的安全性，但由于磁盘是在虚拟机中模拟出来的，所以对其读写速度的改善可能并不明显。接下来介绍软 RAID 5 损坏后的处理方法，这里假设/dev/sdd 已经损坏。

（1）使用"mdadm"命令将已经损坏的/dev/sdd 移除，如下所示。

```
root@ubuntu:~# mdadm /dev/md0 --fail /dev/sdd
mdadm: set /dev/sdd faulty in /dev/md0
```

（2）更换磁盘，将其作为新的软 RAID 5，并将其命名为"/dev/sde"，如下所示。

```
root@ubuntu:~# mdadm /dev/md0 --add /dev/sde
```

```
mdadm: added /dev/sde
```

(3) 查看软 RAID 5 的具体情况，如下所示。

```
root@ubuntu:~# mdadm --detail /dev/md0
/dev/md0:
......                                    //此处省略部分内容
    Number   Major   Minor   RaidDevice   State
       0       8       16        0        active sync   /dev/sdb
       1       8       32        1        active sync   /dev/sdc
       3       8       64        3        spare rebuilding   /dev/sde
//这里损坏的/dev/sdd 已经被成功替换成/dev/sde
```

任务 33 LVM 管理

任务目标

1. 理解 LVM 的工作原理。
2. 了解 LVM 的构成。
3. 实现 LVM 的配置和管理方法。

任务描述

某公司购置了 Linux 服务器,已经部署了软 RAID,现在想要修改磁盘分区的大小,但不能破坏原有的数据。网络管理员小赵决定使用 LVM(Logical Volume Manager,逻辑卷管理器)实现。

任务要求

LVM 是 Linux 操作系统对磁盘分区进行管理的一种机制,可以解决磁盘设备在创建分区后不易修改分区大小的问题。LVM 允许用户对磁盘资源进行动态调整。当用户想要随着实际需求的变化而调整磁盘分区的大小时,使用 LVM 是不错的选择,具体要求如下。

(1)分别添加大小为 8GB、9GB 和 10GB 的三块磁盘,并将这三块磁盘分别创建为物理卷。

(2)将大小为 8GB 和 9GB 的物理卷加入 vg1 卷组中。

(3)在 vg1 卷组中创建一个大小为 6GB 的 lv-1 逻辑卷,使用剩余空间创建一个 lv-2 逻辑卷。

(4)将大小为 10GB 的物理卷扩展到 vg1 卷组中。

(5)为 lv-1 逻辑卷创建 ext4 文件系统,挂载到/disk/lv-1。

(6)为 lv-2 逻辑卷创建 XFS 文件系统,挂载到/disk/lv-2。

任务实施

步骤 1：为虚拟机添加磁盘。

为虚拟机添加三块磁盘的过程参考任务 31。

步骤 2：创建物理卷。

（1）使用"fdisk"命令查看磁盘信息，如下所示。

```
root@ubuntu:~# fdisk -l
Disk /dev/loop0: 111.95 MiB, 117387264 bytes, 229272 sectors
Units: sectors of 1 * 512 = 512 bytes
Sector size (logical/physical): 512 bytes / 512 bytes
……                                             //此处省略部分内容
Disk /dev/sdb: 8 GiB, 8589934592 bytes, 16777216 sectors
Disk model: VMware Virtual S
Units: sectors of 1 * 512 = 512 bytes
Sector size (logical/physical): 512 bytes / 512 bytes
I/O size (minimum/optimal): 512 bytes / 512 bytes

Disk /dev/sdd: 10 GiB, 10737418240 bytes, 20971520 sectors
Disk model: VMware Virtual S
Units: sectors of 1 * 512 = 512 bytes
Sector size (logical/physical): 512 bytes / 512 bytes
I/O size (minimum/optimal): 512 bytes / 512 bytes

Disk /dev/sdc: 9 GiB, 9663676416 bytes, 18874368 sectors
Disk model: VMware Virtual S
Units: sectors of 1 * 512 = 512 bytes
Sector size (logical/physical): 512 bytes / 512 bytes
I/O size (minimum/optimal): 512 bytes / 512 bytes
//可以看出/dev/sdb、/dev/sdc、/dev/sdd 是新添加的硬盘，是没有经过分区和格式化的
```

（2）使用"fdisk"命令，将/dev/sdb、/dev/sdc、/dev/sdd 这三块磁盘创建成 LVM 类型的分区。这里以/dev/sdb 磁盘为例进行介绍，如下所示。

```
root@ubuntu:~# fdisk /dev/sdb

   ……                                          //此处省略部分内容


Command (m for help): n
Partition type
   p   primary (0 primary, 0 extended, 4 free)
   e   extended (container for logical partitions)
Select (default p): p
```

```
Partition number (1-4, default 1): 1
First sector (2048-16777215, default 2048):
Last sector, +/-sectors or +/-size{K,M,G,T,P} (2048-16777215, default 16777215):
Created a new partition 1 of type 'Linux' and of size 8 GiB.


Command (m for help): t
Selected partition 1
Hex code or alias (type L to list all): 8e
Changed type of partition 'Linux' to 'Linux LVM'.

Command (m for help): p
Disk /dev/sdd: 8 GiB, 8589934592 bytes, 16777216 sectors
Disk model: VMware Virtual S
Units: sectors of 1 * 512 = 512 bytes
Sector size (logical/physical): 512 bytes / 512 bytes
I/O size (minimum/optimal): 512 bytes / 512 bytes
Disklabel type: dos
Disk identifier: 0x57ac4b08


Device     Boot Start      End  Sectors Size Id Type
/dev/sdd1       2048  16777215 16775168   8G 8e Linux LVM

Command (m for help): w
The partition table has been altered.
Calling ioctl() to re-read partition table.
Syncing disks.
//可看到分区已经从 Linux 变为 LVM
```

（3）创建物理卷。

将这三块磁盘创建为物理卷，如下所示。

```
root@ubuntu:~# pvcreate /dev/sdb1
  Physical volume "/dev/sdb1" successfully created.
root@ubuntu:~# pvcreate /dev/sdc1
  Physical volume "/dev/sdc1" successfully created.
root@ubuntu:~# pvcreate /dev/sdd1
  Physical volume "/dev/sdd1" successfully created.
```

步骤 3：创建卷组。

在创建好物理卷后，可以使用"vgcreate"命令创建卷组。卷组中可以有多个物理卷，也可以只有一个物理卷。

将/dev/sdb1 和/dev/sdc1 物理卷加入 vg1 卷组中，如下所示。

```
root@ubuntu:~# vgcreate vg1 /dev/sdb1 /dev/sdc1
  Volume group "vg1" successfully created
```

步骤 4：创建逻辑卷。

在创建好卷组后，可以使用"lvcreate"命令创建逻辑卷。

在 vg1 卷组中创建一个大小为 6GB 的 lv-1 逻辑卷，剩余空间创建一个 lv-2 逻辑卷，如下所示。

```
root@ubuntu:~# lvcreate -L 6G -n lv-1 vg1
  Logical volume "lv-1" created.
root@ubuntu:~# lvcreate -L 10.99G -n lv-2 vg1
  Rounding up size to full physical extent 10.99 GiB
  Logical volume "lv-2" created.
```

步骤 5：扩展逻辑卷。

（1）将/dev/sdd1 物理卷扩展到 vg1 卷组中，如下所示。

```
root@ubuntu:~# vgextend vg1 /dev/sdd1
  Volume group "vg1" successfully extended
```

（2）将 lv-1 逻辑卷的大小增加到 12GB，如下所示。

```
root@ubuntu:~# lvextend -L +6G /dev/vg1/lv-1
  Size of logical volume vg1/lv-1 changed from 6.00 GiB (1536 extents) to 12.00 GiB (3072 extents).
  Logical volume vg1/lv-1 successfully resized.
```

（3）为 lv-1 逻辑卷创建 ext4 文件系统，挂载到/disk/lv-1，如下所示。

```
root@ubuntu:~# mkfs.ext4 /dev/vg1/lv-1
mke2fs 1.46.5 (30-Dec-2021)
Creating filesystem with 3145728 4k blocks and 786432 inodes
Filesystem UUID: 491191be-193c-43ad-be2c-f173830ae83e
Superblock backups stored on blocks:
32768, 98304, 163840, 229376, 294912, 819200, 884736, 1605632, 2654208

Allocating group tables: done
Writing inode tables: done
Creating journal (16384 blocks): done
Writing superblocks and filesystem accounting information: done
root@ubuntu:~# mkdir -p /disk/lv-1
root@ubuntu:~# mount /dev/vg1/lv-1 /disk/lv-1
```

步骤 6：压缩逻辑卷。

（1）将 lv-2 逻辑卷的大小减少为 6.99GB，如下所示。

```
root@ubuntu:~# lvreduce -L -4G /dev/vg1/lv-2
  WARNING: Reducing active logical volume to 6.99 GiB.
  THIS MAY DESTROY YOUR DATA (filesystem etc.)
Do you really want to reduce vg1/lv-2? [y/n]: y
```

```
    Size of logical volume vg1/lv-2 changed from 10.99 GiB (2814 extents) to 6.99 GiB
(179 extents)
    Logical volume vg1/lv-2 successfully resized.
```

（2）为 lv-2 逻辑卷创建 XFS 文件系统，挂载到/disk/lv-2，如下所示。

```
root@ubuntu:~# mkfs.xfs /dev/vg1/lv-2
meta-data=/dev/vg1/lv-2          isize=512    agcount=4, agsize=458240 blks
         =                       sectsz=512   attr=2, projid32bit=1
         =                       crc=1        finobt=1, sparse=1, rmapbt=0
         =                       reflink=1    bigtime=0 inobtcount=0
data     =                       bsize=4096   blocks=1832960, imaxpct=25
         =                       sunit=0      swidth=0 blks
Naming   =version 2              bsize=4096   ascii-ci=0, ftype=1
Log      =internal log           bsize=4096   blocks=2560, version=2
         =                       sectsz=512   sunit=0 blks, lazy-count=1
Realtime =none                   extsz=4096   blocks=0, rtextents=0
root@ubuntu:~# mkdir -p /disk/lv-2
root@ubuntu:~# mount /dev/vg1/lv-2 /disk/lv-2
```

任务验收

（1）使用"pvscan"命令查看系统中的物理卷，如下所示。

```
root@ubuntu:~# pvscan                //查看当前系统中的物理卷，与"pvs"和"pvdisplay"命令功能相同
  PV /dev/sda3   VG ubuntu-vg       lvm2 [<28.00 GiB / 14.00 GiB free]
  PV /dev/sdb1                      lvm2 [<8.00 GiB]
  PV /dev/sdc1                      lvm2 [<9.00 GiB]
  PV /dev/sdd1                      lvm2 [<10.00 GiB]
  Total: 4 [54.99 GiB] / in use: 1 [<28.00 GiB] / in no VG: 3 [<27.00 GiB]
```

（2）使用"vgs"命令查看当前系统中 vg1 卷组的信息，如下所示。

```
root@ubuntu:~# vgs vg1    //查看当前系统中 vg1 卷组的信息
  VG  #PV #LV #SN Attr   VSize  VFree
  vg1   2   0   0 wz--n- 16.99g 16.99g
```

（3）使用"lvdisplay"命令查看 lv-1 逻辑卷的信息，如下所示。

```
root@ubuntu:~# lvdisplay /dev/vg1/lv-1
  --- Logical volume ---
  LV Path                /dev/vg1/lv-1
  LV Name                lv-1
  VG Name                vg1
  LV UUID                3zTqqT-GYHu-KBLV-72KU-d1RZ-mdCj-uk0NFC
  LV Write Access        read/write
  LV Creation host, time ubuntu, 2024-01-21 07:13:41 +0000
  LV Status              available
```

```
# open                  0
LV Size                 6.00 GiB
Current LE              1536
Segments                1
Allocation              inherit
Read ahead sectors      auto
- currently set to      256
Block device            253:1
```

(4)使用"df"命令查看 lv-1 和 lv-2 逻辑卷的使用情况,如下所示。

```
root@ubuntu:~# df -TH|grep lv
/dev/mapper/ubuntu--vg-ubuntu--lv   ext4    15G   6.2G  7.9G   44%  /
/dev/mapper/vg1-lv--1               ext4    13G   25k   12G    1%   /disk/lv-1
/dev/mapper/vg1-lv--2               xfs     7.5G  87M   7.5G   2%   /disk/lv-2
```

任务 34　配置 DNS 服务器

任务目标

1．了解 DNS 服务器的工作原理。
2．实现 DNS 服务器的配置和验证。

任务描述

为了让公司员工能够简单、快捷地访问本地网络和互联网上的资源，以及方便公司向外发布网站，就需要在公司局域网上配置 DNS 服务器。公司将此任务交给网络管理员小赵。

任务要求

通过安装 DNS 服务器，并配置其配置文件和正向解析区域文件、反向解析区域文件，为用户提供 DNS 服务。配置 DNS 服务器的网络拓扑结构，如图 34.1 所示。

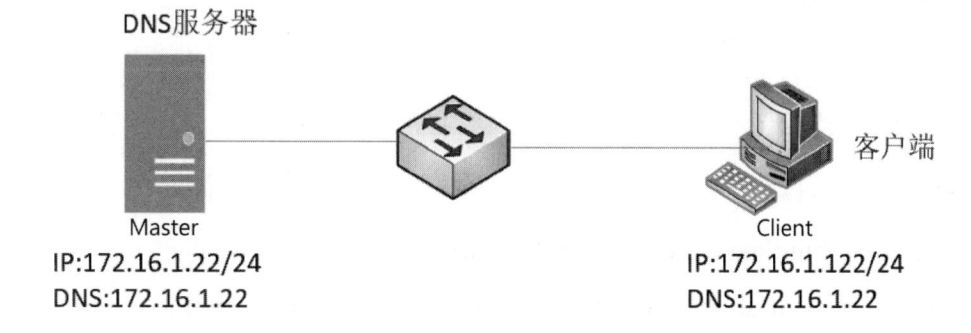

图 34.1　配置 DNS 服务器的网络拓扑结构

具体要求如下。
（1）虚拟机的网络配置方式统一为仅主机模式。
（2）配置 DNS 服务器，具体要求如表 34.1 所示。

表 34.1 配置 DNS 服务器的具体要求

项 目	说 明
DNS 服务器 IP 地址	172.16.1.22/24
域名	phei.com.cn
正向解析区域文件	db.phei.com.cn.zone
反向解析区域文件	db.172.16.1.zone
SOA、NS 资源记录	默认值
MX 资源记录	mail.phei.com.cn
A 资源记录	master.phei.com.cn(172.16.1.22) mail.phei.com.cn(172.16.1.23) client.phei.com.cn(172.16.1.122)
CNAME 资源记录	设置主机 master 名称为"www"
转发器	8.8.8.8

（3）在反向解析区域文件中添加与正向解析区域文件对应的 PTR 记录。

（4）在客户端上测试 DNS 服务器的正确性。

任务实施

步骤 1：配置 DNS 服务器的 IP 地址等信息，在前面的任务中已经介绍，这里不再详述。

步骤 2：安装 bind9 软件包。使用"apt install -y bind9"命令安装 DNS 服务器需要的 bind9 软件包，如下所示。

```
root@master:~# apt install -y bind9                    //安装bind9软件包
root@ubuntu:~# apt policy bind9                        //查询是否安装
bind9:
  Installed: 1:9.18.18-0ubuntu0.22.04.1
  Candidate: 1:9.18.18-0ubuntu0.22.04.1
  Version table:
 *** 1:9.18.18-0ubuntu0.22.04.1 500
……                                                    //此处省略部分内容
//ii 表示已安装成功
```

步骤 3：修改/etc/bind/named.conf.default-zones 主配置文件。

在/etc/bind/named.conf.default-zones 主配置文件的末尾添加如下内容。

```
root@master:~# vim /etc/bind/named.conf.default-zones   //在主配置文件的末尾添加内容
zone "phei.com.cn" IN {
        type master;
        file "/etc/bind/db.phei.com.cn.zone";
};
zone "1.16.172.in-addr.arpa" IN {
        type master;
```

```
        file "/etc/bind/db.172.16.1.zone";
};
```

步骤 4：创建正向解析区域文件和反向解析区域文件。

在/etc/bind 目录下创建正向解析区域文件 db.phei.com.cn.zone 和反向解析区域文件 db.172.16.1.zone，如下所示。

```
root@master:~# cd /etc/bind
root@master:/etc/bind# cp db.local db.phei.com.cn.zone
root@master:/etc/bind# cp db.127 db.172.16.1.zone
root@master:/etc/bind# ls -l *zone
-rw-r--r-- 1 root bind 271 Dec  25 17:42 db.172.16.1.zone
-rw-r--r-- 1 root bind 281 Dec  25 16:14 db.phei.com.cn.zone
```

步骤 5：配置正向解析区域文件。

打开 DNS 服务器的/etc/bind 目录下的正向解析区域文件 db.phei.com.cn.zone 并进行配置，配置后的内容如下所示。

```
root@master:/etc/bind# vim db.phei.com.cn.zone
;
; BIND data file for local loopback interface
;
$TTL    604800
@       IN      SOA     localhost. root.localhost. (
                              2         ; Serial
                         604800         ; Refresh
                          86400         ; Retry
                        2419200         ; Expire
                         604800 )       ; Negative Cache TTL
;
@       IN      NS      master.
@       IN      MX      5       mail.phei.com.cn.
master  IN      A       172.16.1.22
mail    IN      A       172.16.1.23
client  IN      A       172.16.1.122
www     IN      CNAME   master
```

步骤 6：配置反向解析区域文件。

打开 DNS 服务器的/etc/bind 目录下的反向解析区域文件 db.172.16.1.zone 并进行配置，配置后的内容如下所示。

```
root@master:/etc/bind# vim db.172.16.1.zone
;
; BIND reverse data file for local loopback interface
;
$TTL    604800
```

```
@      IN     SOA    localhost. root.localhost. (
                            1           ; Serial
                       604800           ; Refresh
                        86400           ; Retry
                      2419200           ; Expire
                       604800 )         ; Negative Cache TTL
;
@      IN     NS     master.
@      IN     MX     5      mail.phei.com.cn.
22     IN     PTR    master.phei.com.cn.
23     IN     PTR    mail.phei.com.cn.
122    IN     PTR    client.phei.com.cn.
```

知识链接:

正向解析区域文件和反向解析区域文件常用的参数及其功能如表34.2所示。

表34.2 正向解析区域文件和反向解析区域文件常用的参数及其功能

参 数	功 能
$TTL 1D	表示资源记录的生存周期（Time To Live，TTL），代表地址解析记录的默认缓存天数。单位为秒，这里的"1D"表示1天
@	表示当前DNS的区域名，如"phei.com.cn."
IN	表示将当前记录标识为一个INTERNET的DNS资源记录
SOA	表示起始授权记录（Start Of Authority，SOA），用于区分资源记录的类型。常见的资源记录类型有SOA、NS（Name Server，域名服务器）
rname.invalid.	表示区域管理员的邮件地址
serial	表示本区域文件的版本号或更新序列号，当从DNS服务器要进行数据同步时，会比较这个号码，如果发现主服务器的序号比自己的大，则进行更新，否则忽略。一般来说会使用容易记忆的数字。例如，使用时间作为序号，即将2022年8月12日第1个版本的版本号写作"2022081201"
refresh	表示刷新时间。从DNS服务器根据定义的时间，周期性地检查主DNS服务器的序列号是否发生了变化。如果发生了变化，则更新自己的区域文件。这里的"1D"表示1天
retry	表示从DNS服务器在同步失败后，进行重试的时间间隔。这里的"1H"表示1小时
expiry	表示过期时间。如果从DNS服务器在有效期内无法与主DNS服务器取得联系，则从DNS服务器不再响应查询请求，无法对外提供域名解析服务。这里的"1W"表示1周
minimum	表示对于没有特别指定存活周期的资源记录默认取minimum的值为1天，即86 400秒。这里的"3H"表示3小时
NS	表示资源记录（Name Server，NS）。资源记录指定该域名由哪个DNS服务器进行解析，格式为"@ IN NS master.phei.com.cn."

续表

参　　数	功　　能
A 和 AAAA 资源记录	表示域名与 IP 地址的映射关系。A 资源记录为 IPv4 地址，AAAA 资源记录为 IPv6 地址，格式为"master IN A 192.168.1.201"
CNAME	表示别名记录，格式为"www1 IN CNAME www"，这里"www1"表示 www 主机名称
MX	定义邮件服务器，优先级默认为 10，数字越小，优先级越高，格式为"@ IN MX 10 mail.yiteng.cn."
PTR	指针记录（Pointer），表示 IP 地址与域名的映射关系，反向解析区域文件与正向解析区域文件主要区别在此指针记录上。指针记录常被用于用户 DNS 的反向解析，格式为"201 IN PTR www.phei.com.cn."，这里"201"表示 IP 地址中的主机号，IP 地址为 192.168.1.201，完整的记录名为"201.1.168.192.in-addr.arpa"

步骤 7：配置转发器。

当本地 DNS 服务器无法对 DNS 客户端的解析请求进行解析时，本地 DNS 服务器将转发 DNS 客户端的解析请求到上游 DNS 服务器。此时本地 DNS 服务器又称转发服务器，而上游 DNS 服务器又称转发器。配置转发器的方法如下所示。

```
root@ubuntu:~# vim /etc/bind/named.conf.options
options {
……                                          //省略部分代码
        forwarders {
            8.8.8.8;                         //转发到 IP 地址为 8.8.8.8 的上游 DNS 服务器
        };

……                                          //省略部分代码
};
```

步骤 8：重启 DNS 服务。

在转发器配置完成后，使用"systemctl restart named"命令重启 DNS 服务，并使用"systemctl enable named"命令设置开机自启动，如下所示。

```
root@master:~# systemctl restart named
root@master:~# systemctl enable named
```

步骤 9：配置 DNS 客户端。

配置 DNS 客户端，要确保两台主机之间的网络连接正常。客户端的 DNS 服务配置如下所示。

```
root@client:~# cat /etc/resolv.conf
nameserver 172.16.1.22
```

任务验收

在 DNS 客户端上测试 DNS 服务是否配置正确。

bind9 软件包提供了 3 个实用的 DNS 测试工具——nslookup、dig 和 host。dig 和 host 是命令行工具，而 nslookup 工具有命令行模式和交互模式。

打开 nslookup 工具，在命令行模式下输入"nslookup"，并按"Enter"键，切换至交互模式。在 nslookup 工具的交互模式下进行 DNS 服务配置的测试，如下所示。

```
root@client:~# nslookup
> master.phei.com.cn                              //正向解析
Server:         172.16.1.22                       //显示 DNS 服务器的 IP 地址
Address:        172.16.1.22#53
Name:   master.phei.com.cn
Address: 172.16.1.22
> 172.16.1.22                                     //反向解析
22.1.16.172.in-addr.arpa   name = master.phei.com.cn.1.16.172.in-addr.arpa
> set type=NS                                     //查询区域的 DNS 服务器
> phei.com.cn                                     //输入域名
Server:         172.16.1.22
Address:        172.16.1.22#53
phei.com.cn     nameserver = master.
> set type=MX                                     //查询区域的邮件服务器
> phei.com.cn                                     //输入域名
Server:         172.16.1.22
Address:        172.16.1.22#53
phei.com.cn    mail exchanger = 5 mail.phei.com.cn.
> set type=CNAME                                  //查询别名
> www.phei.com.cn                                 //输入域名
Server:         172.16.1.22
Address:        172.16.1.22#53
www.phei.com.cn  canonical name = master.phei.com.cn.
> exit                                            //退出
```

任务 35　配置 DHCP 服务器

任务目标

1．了解 DHCP 服务器的工作原理。
2．正确安装、配置和启动 DHCP 服务器。
3．让 DHCP 客户端正确获取服务器的 IP 地址。

任务描述

最近一段时间，某公司的网络管理员小赵收到了不少计算机出现 IP 地址冲突问题的求助。经检查发现，这些问题是部分员工自行设置 IP 地址造成的。小赵需要解决这一问题。于是小赵准备在信息中心的 Linux 服务器上安装 DHCP 软件包，并将其配置为 DHCP 服务器，使用动态分配 IP 地址的方式解决 IP 地址冲突的问题。

任务要求

在信息中心的 Linux 服务器上安装 DHCP 软件包并将其配置为 DHCP 服务器，从而实现动态分配 IP 地址的功能。DHCP 服务器可以为主机动态分配 IP 地址，能很好地解决 IP 地址冲突的问题。配置 DHCP 服务器的网络拓扑结构如图 35.1 所示。

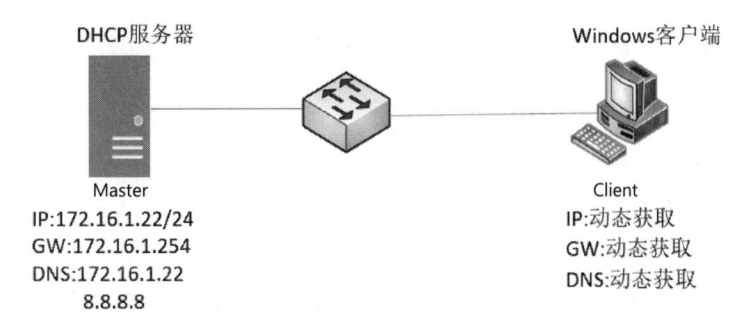

图 35.1　配置 DHCP 服务器的网络拓扑结构

具体要求如下。

（1）在 Master 上，配置 DHCP 服务器为内部计算机，并为其分配 IP 地址，IP 地址为 172.16.1.22/24。

（2）分配的地址池为 172.16.1.2~172.16.1.21，子网掩码为 24 位。

（3）网关地址为 172.16.1.254，DNS 服务器的地址为 172.16.1.22 和 8.8.8.8。

（4）在 Client 上，动态获取 IP 地址等信息。

任务实施

步骤 1：设置 Master 的 IP 地址等信息，在前面的任务中已经介绍，这里不再赘述。

步骤 2：安装 DHCP 软件包。使用 "apt install -y isc-dhcp-server" 命令安装 DHCP 服务器需要的 DHCP 软件包，如下所示。

```
root@master:~# apt install -y isc-dhcp-server
root@master:~# dpkg -l isc-dhcp-server
Desired=Unknown/Install/Remove/Purge/Hold
| Status=Not/Inst/Conf-files/Unpacked/halF-conf/Half-inst/trig-aWait/Trig-pend
|/ Err?=(none)/Reinst-required (Status,Err: uppercase=bad)
||/ Name           Version              Architecture Description
+++-==============-====================-============-===========================
ii  isc-dhcp-server 4.4.1-2.3ubuntu2.4  amd64        ISC DHCP server for automatic IP address assignment
//ii 表示已安装成功
```

步骤 3：修改/etc/dhcp/dhcpd.conf 主配置文件，修改完成后保存退出，具体内容如下所示。

```
root@master:~# vim /etc/dhcp/dhcpd.conf
default-lease-time 600;
max-lease-time 7200;
ddns-update-style none;
subnet 172.16.1.0 netmask 255.255.255.0 {
  range 172.16.1.2 172.16.1.21;
  option domain-name-servers 172.16.1.22,8.8.8.8;
  option routers 172.16.1.254;
    option broadcast-address 172.16.1.255;
  default-lease-time 600;
  max-lease-time 7200;
}
```

/etc/dhcp/dhcpd.conf 主配置文件由参数、选项和声明 3 种要素组成。

（1）参数：表明如何执行任务，是否要执行任务，格式是"参数名 参数值;"。DHCP 服务器常用的参数及其功能如表 35.1 所示。

表 35.1　DHCP 服务器常用的参数及其功能

参　　数	功　　能
ddns-update-style	设置 DNS 服务器动态更新的类型
default-lease-time	默认租约时间，单位是秒
max-lease-time	最大租约时间，单位是秒
log-facility	指定日志文件名称
hardware	指定网卡接口类型和 MAC 地址
server-name	指定 DHCP 服务器的名称
fixed-address	分配给 DHCP 客户端一个固定的 IP 地址

（2）选项：通常用来配置 DHCP 客户端的可选参数，以"option"关键字为开始，如"option 参数名 参数值;"。DHCP 服务器常用的选项及其功能如表 35.2 所示。

表 35.2　DHCP 服务器常用的选项及其功能

选　　项	功　　能
subnet-mask	为 DHCP 客户端设置子网掩码
domain-name	为 DHCP 客户端指定 DNS 服务域名
domain-name-servers	为 DHCP 客户端指定 DNS 服务器地址
host-name	为 DHCP 客户端指定主机名称
routers	为 DHCP 客户端设置默认网关
broadcast-address	为 DHCP 客户端设置广播地址

（3）声明：一般用来指定 IP 作用域，定义 DHCP 客户端分配的 IP 地址等。两种最常用的声明是 subnet 声明和 host 声明。subnet 声明用于定义作用域和指定子网；host 声明用于定义保留地址，实现 IP 地址和 DHCP 客户端 MAC 地址的绑定。DHCP 服务器常用的声明及其功能如表 35.3 所示。

表 35.3　DHCP 服务器常用的声明及其功能

声　　明	功　　能
shared-network	告知是否允许子网络分享相同网络
subnet	定义作用域并指定子网
range	IP 地址范围
host	定义保留地址

步骤 4：重启 DHCP 服务。

在配置完成后，重启 DHCP 服务，并设置开机自动启动，如下所示。

```
root@master:~# systemctl restart isc-dhcp-server
root@master:~# systemctl enable isc-dhcp-server
```

任务验收

（1）配置 Windows 客户端。将 Windows 客户端配置为 DHCP 客户端比较简单，可以采用图形化配置，配置步骤如下。

右击桌面上的"网上邻居"图标，在弹出的快捷菜单中选择"属性"命令，打开"网络连接"窗口，右击"本地连接"图标，打开"本地连接属性"对话框，双击"Internet 协议(TCP/IPv4)"选项，打开"Internet 协议版本 4(TCP/IPv4)属性"对话框，如图 35.2 所示。

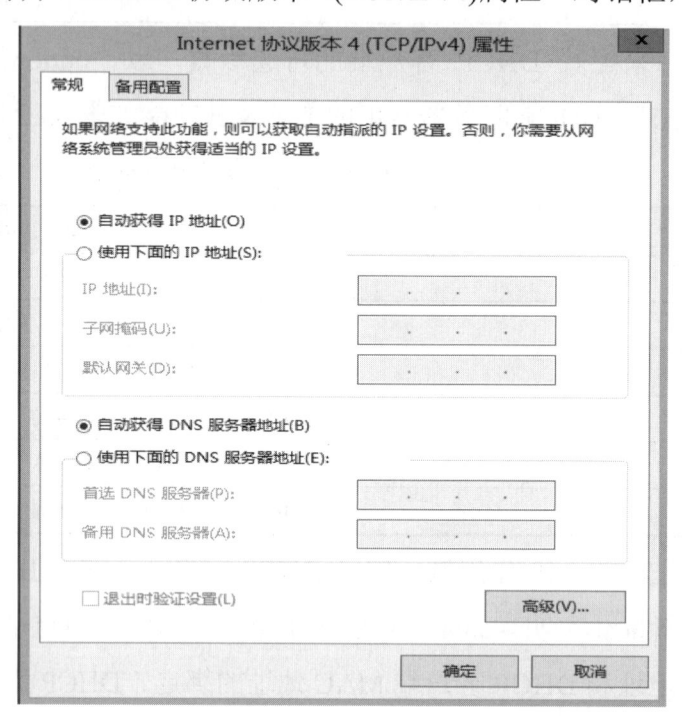

图 35.2 "Internet 协议版本 4(TCP/IPv4)属性"对话框

选中"自动获得 IP 地址"和"自动获得 DNS 服务器地址"单选按钮，单击"确定"按钮即可完成客户端的配置。

在虚拟机界面的菜单栏中，选择"编辑"→"虚拟网络编辑器"命令，打开"虚拟网络编辑器"对话框，如图 35.3 所示，取消勾选"使用本地 DHCP 服务将 IP 地址分配给虚拟机"复选框。

在计算机桌面上选择"开始"→"运行"命令，在打开的"运行"窗口中，输入"cmd"命令并按"Enter"键。在打开的命令提示符窗口中，可以通过"ipconfig/release"命令释放获得的 IP 地址，使用"ipconfig/renew"命令重新获得 IP 地址。通过"ipconfig/all"命令查看获得的 IP 地址参数，如图 35.4 所示，可以看出 DHCP 客户端已经成功获得 IP 地址。

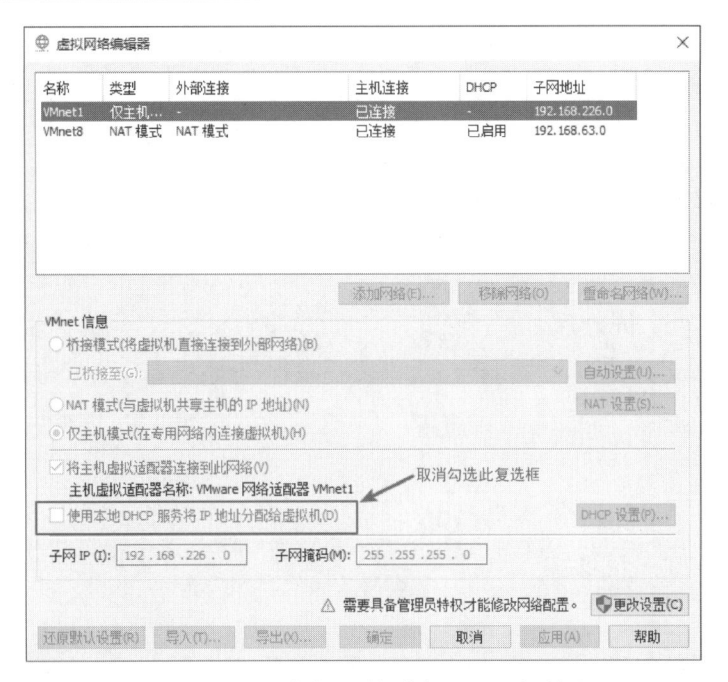

图 35.3 "虚拟网络编辑器"对话框

图 35.4 查看 DHCP 客户端获得的 IP 地址参数

（2）在 DHCP 服务器上查看 IP 地址信息，如下所示。

```
root@ubuntu:/etc/dhcp# grep -v '#' /var/lib/dhcp/dhcpd.leases
authoring-byte-order little-endian;
server-duid "\000\001\000\001-E%2\000\014)3\206d";
lease 172.16.1.2 {                    //DHCP 客户端获取的 IP 地址
  starts 4 2024/01/25 13:58:42;
  ends 4 2024/01/25 14:08:42;
```

```
    cltt 4 2024/01/25 13:58:42;
    binding state active;
    next binding state free;
    rewind binding state free;
    hardware ethernet 00:0c:29:91:86:83;
    uid "\001\000\014)\221\206\203";
    set vendor-class-identifier = "MSFT 5.0";
    client-hostname "client";
}
```

任务 36　配置 Apache 服务器

任务目标

1．使用 Apache 常见配置项进行配置。
2．实现基于域名和端口号的虚拟主机的配置。

任务描述

某公司的网络管理员小赵根据公司的业务需求，要将公司程序员开发好的网站部署到信息中心的 Web 服务器上。公司使用的是 Ubuntu 服务器，因此现在需要安装 Apache 软件包，并将其配置为 Apache 服务器。

任务要求

在信息中心的 Ubuntu 服务器上安装 Apache 软件包，并将其配置为 Apache 服务器，从而实现网站部署的功能。Ubuntu 服务器安装 Apache 软件包后支持在同一台服务器上发布多个网站，这些网站也被称为虚拟主机，这些网站在 IP 地址、端口、主机名中至少有一项与其他网站不同。配置 Apache 服务器的网络拓扑结构，如图 36.1 所示。

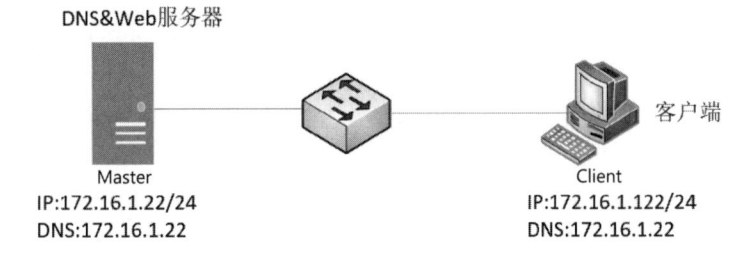

图 36.1　配置 Apache 服务器的网络拓扑结构

具体要求如下。

（1）在 Master 上，配置 DNS 服务器，设置 IP 地址为 172.16.1.22/24，域名为"phei.com.cn"。

（2）配置 Apache 虚拟主机如表 36.1 所示。

表 36.1　配置 Apache 虚拟主机

项　　目	说　　明
基于域名的虚拟主机	IP 地址：172.16.1.22/24 DNS 地址：172.16.1.22 域名：www1.phei.com.cn 和 www2.phei.com.cn 根目录：/vh/www1 和 /vh/www2 首页内容："This is www1 homepage." 和 "This is www2 homepage." 首页文件：index.html
基于端口的虚拟主机	IP 地址：172.16.1.22/24 DNS 地址：172.16.1.22 端口号：www.phei.com.cn:8088 和 www.phei.com.cn:8089 根目录：/vh/8088 和 /vh/8089 首页内容："This is 8088 homepage." 和 "This is 8089 homepage." 首页文件：index.html

（3）在 Client 上，使用 "curl" 命令测试 Apache 服务器配置的正确性。

任务实施

步骤 1：配置基于域名的虚拟主机。

（1）在 Ubuntu 服务器的正向解析区域文件中添加两条 CNAME 资源记录，如下所示。

```
root@master:~# vim /etc/bind/db.phei.com.cn.zone
www1    IN      CNAME     master
www2    IN      CNAME     master
```

（2）安装 Apache 软件包。使用 "apt install -y apache2" 命令安装 Apache 服务器需要的软件包，如下所示。

```
root@master:~# apt install -y apache2
root@master:~# dpkg -l apache2
Desired=Unknown/Install/Remove/Purge/Hold
| Status=Not/Inst/Conf-files/Unpacked/halF-conf/Half-inst/trig-aWait/Trig-pend
|/ Err?=(none)/Reinst-required (Status,Err: uppercase=bad)
||/ Name           Version              Architecture Description
+++-==============-====================-============-================================
ii  apache2        2.4.52-1ubuntu4.7    amd64        Apache HTTP Server
//ii 表示已安装成功
```

（3）分别为两个网站创建文档根目录和首页文件，如下所示。

```
root@master:~# mkdir -p /vh/www1
root@master:~# mkdir -p /vh/www2
```

```
root@master:~# echo "This is www1 homepage.">/vh/www1/index.html
root@master:~# echo "This is www2 homepage.">/vh/www2/index.html
```

（4）新建与虚拟主机对应的/etc/apache2/sites-available/vhost1.conf 配置文件，为两台虚拟主机分别指定文件根目录，如下所示。

```
root@master:~# cd /etc/apache2/sites-available/
root@master:/etc/apache2/sites-available# cp 000-default.conf vhost1.conf
root@master:/etc/apache2/sites-available# vim vhost1.conf
    <VirtualHost 172.16.1.22:80>
        ServerName www1.phei.com.cn
        ServerAdmin webmaster@phei.com.cn
        DocumentRoot /vh/www1
        <Directory /vh/www1>
            AllowOverride none
            Require all granted
        </Directory>
    </VirtualHost>
    <VirtualHost 172.16.1.22:80>
        ServerName www2.phei.com.cn
        ServerAdmin webmaster@phei.com.cn
        DocumentRoot /vh/www2
        <Directory /vh/www2>
            AllowOverride none
            Require all granted
        </Directory>
    </VirtualHost>
```

知识链接：

在 Apache 服务器的主配置文件和虚拟主机配置文件中都需要使用 Directory 配置段。<Directory>和</Directory>是一对标签，标签仅对指定的目录有效。Directory 配置段包含的选项及其功能如表 36.2 所示。

表 36.2　Directory 配置段包含的选项及其功能

选　　项	功　　能
Options	配置指定目录具体使用的功能特性
AllowOverride	设置是否将后缀为".htaccess"的文件作为配置文件，可以允许该文件的全部命令，也可以只允许某些类型的指定，或者全部禁止
Order	控制默认访问状态，以及 Allow 和 Deny 选项指定的生效顺序
Allow	控制可以访问的主机。可以根据主机名、IP 地址、IP 范围或其他环境变量的定义进行控制
Deny	限制访问 Apache 服务器的主机列表，其语法和参数与 Allow 选项完全相同

（5）只有使用"a2ensite"命令激活 vhost1.conf 文件中的配置，Web 服务器的站点内容

才能正常显示，如下所示。

```
root@master:/etc/apache2/sites-available#a2ensite vhost1.conf
Enabling site vhost1.
To activate the new configuration, you need to run:
  systemctl reload apache2
```

知识链接：

"a2ensite"是 Apache 服务的快速切换命令之一。只有使用"a2ensite"命令来激活配置文件的配置，服务器的站点内容才能正常显示。常用的 Apache 服务的快速切换命令及其功能如表 36.3 所示。

表 36.3 常用的 Apache 服务的快速切换命令及其功能

命　　令	功　　能
a2ensite	激活/etc/apache2/sites-available 中包含配置文件的站点
a2dissite	禁用/etc/apache2/sites-available 中包含配置文件的站点
a2enmod	启用 Apache 服务的某个模块
a2dismod	禁用 Apache 服务的某个模块
a2enconf	启用某配置文件
a2disconf	禁用某配置文件

（6）重启 Apache 服务，并设置开机自动启动，如下所示。

```
root@master:~# systemctl restart apache2
root@master:~# systemctl enable apache2
```

步骤 2：配置基于端口的虚拟主机。

（1）在/etc/apache2/ports.conf 配置文件中，添加 8088 和 8089 两个监听端口，如下所示。

```
root@master:~# vim /etc/apache2/ports.conf
Listen 8088
Listen 8089
```

（2）分别为两台虚拟主机创建文档和首页文件，如下所示。

```
root@master:~# mkdir -p /vh/8088
root@master:~# mkdir -p /vh/8089
root@master:~# echo "This is 8088 homepage.">/vh/8088/index.html
root@master:~# echo "This is 8089 homepage.">/vh/8089/index.html
```

（3）新建与虚拟主机对应的/etc/apache2/sites-available/vhost2.conf 配置文件，为两台虚拟主机分别指定文件根目录，如下所示。

```
root@master:~# cd /etc/apache2/sites-available/
root@master:/etc/apache2/sites-available# cp 000-default.conf vhost2.conf
root@master:/etc/apache2/sites-available# vim vhost2.conf
    <Virtualhost 172.16.1.22:8088>
        ServerName      www.phei.com.cn
        DocumentRoot    /vh/8088
        <Directory /vh/8088>
            AllowOverride none
            Require all granted
        </Directory>
    </Virtualhost>
    <Virtualhost 172.16.1.22:8089>
        ServerName      www.phei.com.cn
        DocumentRoot    /vh/8089
        <Directory /vh/8089>
            AllowOverride none
            Require all granted
        </Directory>
</Virtualhost>
```

（4）只有使用"a2ensite"命令激活 vhost2.conf 文件中的配置，Web 服务器的站点内容才能正常显示，如下所示。

```
root@master:/etc/apache2/sites-available#a2ensite vhost2.conf
Enabling site vhost2.
To activate the new configuration, you need to run:
  systemctl reload apache2
```

（5）重启 Apache 服务，并设置开机自动启动，如下所示。

```
root@master:~# systemctl restart apache2
root@master:~# systemctl enable apache2
```

任务验收

（1）在 Client 上，查看 Client 的 DNS 地址指向的服务器，确保两台主机之间的网络连接正常。

```
root@client:~# cat /etc/resolv.conf          //查看Client的DNS地址指向的服务器
nameserver 172.16.1.22
```

（2）在命令提示符窗口中，使用"curl"命令分别测试 Apache 服务器配置的正确性，如下所示。

```
root@client:~# curl http://www1.phei.com.cn
This is www1 homepage.
root@client:~# curl http://www2.phei.com.cn
```

```
This is www2 homepage.
root@client:~# curl http://www.phei.com.cn:8088
This is 8088 homepage.
root@client:~# curl http://www.phei.com.cn:8089
This is 8089 homepage.
```

任务 37　配置 Nginx 服务器

任务目标

1．使用 Nginx 常见配置项的配置。
2．实现基于域名和基于端口的虚拟主机的配置。

任务描述

某公司的网络管理员小赵根据公司的业务需求，要将公司程序员开发好的网站部署到信息中心的 Web 服务器上，并向人们展示。公司使用的是 Linux 服务器，需要安装 Nginx 软件包，并将其配置为 Nginx 服务器。

任务要求

Web 服务器的配置主要是通过修改 Nginx 服务器的配置文件来实现的。配置 Nginx 服务器的网络拓扑结构，如图 37.1 所示。

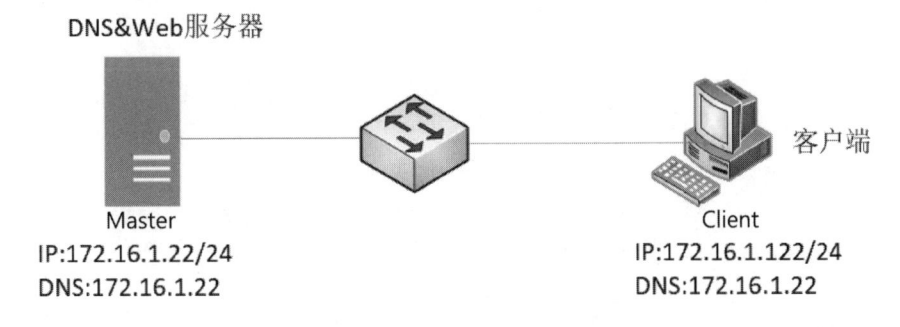

图 37.1　配置 Nginx 服务器的网络拓扑结构

具体要求如下。
（1）在 Master 上，配置 DNS 服务器，设置 IP 地址为 172.16.1.22/24，域名为 "phei.com.cn"。
（2）配置 Nginx 虚拟主机如表 37.1 所示。

表 37.1　配置 Nginx 虚拟主机

项　　目	说　　明
基于域名的虚拟主机	IP 地址：172.16.1.22/24 DNS 地址：172.16.1.22 域名：www1.phei.com.cn 和 www2.phei.com.cn 根目录：/vh/www1 和 /vh/www2 首页内容："This is www1 homepage." 和 "This is www2 homepage." 首页文件：index.html
基于端口的虚拟主机	IP 地址：172.16.1.22/24 DNS 地址：172.16.1.22 端口号：www.phei.com.cn:8088 和 www.phei.com.cn:8089 根目录：/vh/8088 和 /vh/8089 首页内容："This is 8088 homepage." 和 "This is 8089 homepage." 首页文件：index.html

（3）在 Client 上，使用 "curl" 命令测试 Nginx 服务器配置的正确性。

任务实施

步骤 1：配置基于域名的虚拟主机。

（1）在 Linux 服务器的正向解析区域文件中添加两条 CNAME 资源记录，如下所示。

```
root@master:~# vim /etc/bind/db.phei.com.cn.zone
www1    IN    CNAME    master
www2    IN    CNAME    master
```

（2）安装 Nginx 软件包。使用 "apt install -y nginx" 命令安装 Nginx 服务器需要的软件包，如下所示。

```
root@master:~# apt install -y nginx
root@master:~# dpkg -l nginx
Desired=Unknown/Install/Remove/Purge/Hold
| Status=Not/Inst/Conf-files/Unpacked/halF-conf/Half-inst/trig-aWait/Trig-pend
|/ Err?=(none)/Reinst-required (Status,Err: uppercase=bad)
||/ Name         Version            Architecture Description
+++-==================-========================-============-==================================
ii  nginx        1.18.0-6ubuntu14.4 amd64        small, powerful, scalable web/proxy server
//ii 表示已安装成功
```

（3）分别为两个网站创建文档根目录和首页文件，如下所示。

```
root@master:~# mkdir -p /vh/www1
root@master:~# mkdir -p /vh/www2
root@master:~# echo "This is www1 homepage.">/vh/www1/index.html
```

```
root@master:~# echo "This is www2 homepage.">/vh/www2/index.html
```

（4）新建和虚拟主机对应的/etc/nginx/conf.d/vhost.conf 配置文件，为两台虚拟主机分别指定文件根目录，如下所示。

```
root@master:~# vim /etc/nginx/conf.d/vhost.conf
    server {
        listen          80;
        server_name     www1.phei.com.cn;
        root            /vh/www1;
        index           index.html;
        }
    server {
        listen          80;
        server_name     www2.phei.com.cn;
        root            /vh/www2;
        index           index.html;
        }
```

知识链接：

在 nginx.conf 配置文件中包含一些单行的指令和配置段。指令的基本语法格式是"参数名 参数值"，配置段是用一对标签表示的配置选项。Nginx 服务的参数及其功能如表37.2 所示。

表 37.2 Nginx 服务的参数及其功能

参 数	功 能
user nginx	设置以何种身份运行 Nginx 服务，默认为 Nginx 用户身份
worker_processes auto	设置 Nginx 的进程数量，默认为 auto，通常 CPU 有多少个核，就将 worker_processes 的值设置为多少
error_log /var/log/nginx/error.log	错误日志的位置，默认值为/var/log/nginx/error.log
pid /run/nginx.pid	指定保存 Nginx PID 的文件
include /usr/share/nginx/modules/*.conf	包含所启用的模块配置文件
worker_connections 1024	设置单个 worker process 进程的最大并发连接数
sendfile on	启用高效文件传输模式
keepalive_timeout 65	设置客户端保持连接的超时时间（单位为秒），如果超时则中断连接
include /etc/nginx/mime.types	包含文件扩展名和文件类型对应关系文件
default_type application/octet-stream	默认文件类型
server	Nginx 默认的 Web 服务器，也可视为默认的虚拟主机
listen 80 default_server	IPv4 默认端口为 80

续表

参　数	功　能
listen [::]:80 default_server	IPv6 默认端口为 80
server_name _	Web 服务域名
root usr/share/nginx/html	Web 服务的根目录
include /etc/nginx/default.d/*.conf	加载默认配置
location /	定义 404 错误页面
error_page 500 502 503 504 /50X.html	定义 50X 错误页面
# Settings for a TLS enabled server.	SSL 相关配置，默认为注释状态

（5）重启 Nginx 服务，并设置开机自动启动，如下所示。

```
root@master:~# systemctl restart nginx
root@master:~# systemctl enable nginx
```

步骤 2：配置基于端口的虚拟主机。

（1）分别为两台虚拟主机创建文档根目录和首页文件，如下所示。

```
root@master:~# mkdir -p /vh/8088
root@master:~# mkdir -p /vh/8089
root@master:~# echo "This is 8088 homepage.">/vh/8088/index.html
root@master:~# echo "This is 8089 homepage.">/vh/8089/index.html
```

（2）在 /etc/nginx/conf.d/vhost.conf 配置文件中，添加基于端口号的虚拟主机的配置，如下所示。

```
root@master:~# vim /etc/nginx/conf.d/vhost.conf
    ……                                          //此处省略部分内容
    server {
        listen       8088;
        server_name  www.phei.com.cn;
        root         /vh/8088;
        index        index.html;
    }
    server {
        listen       8089;
        server_name  www.phei.com.cn;
        root         /vh/8089;
        index        index.html;
    }
```

（3）重启 Nginx 服务，并设置开机自动启动，如下所示。

```
root@master:~# systemctl restart nginx
root@master:~# systemctl enable nginx
```

任务验收

(1) 在 Client 上，查看 Client 的 DNS 地址指向的服务器，确保两台计算机之间的网络连接正常。

```
root@client:~# cat /etc/resolv.conf         //查看 Client 的 DNS 地址指向的服务器
nameserver 172.16.1.22
```

(2) 在命令提示符窗口中，使用"curl"命令分别测试 Nginx 服务器配置的正确性，如下所示。

```
root@client:~# curl http://www1.phei.com.cn
This is www1 homepage.
root@client:~# curl http://www2.phei.com.cn
This is www2 homepage.
root@client:~# curl http://www.phei.com.cn:8088
This is 8088 homepage.
root@client:~# curl http://www.phei.com.cn:8089
This is 8089 homepage.
```

任务 38　配置证书服务

任务目标

1．理解证书服务的作用。
2．熟练配置证书服务。

任务描述

某公司的 Web 站点已建立完成，但在应用过程中发现有用户信息在通信过程中被泄露。因此，网络管理员小赵决定采用更可靠的 SSL 方式，利用证书服务在一定程度上保证访问 Web 站点的安全性。

任务要求

针对公司的需求，可以使用 SSL（Secure Socket Layer，安全套接字层）协议实现 Web 站点的安全访问。SSL 用于安全地传送数据，集中到每个 Web 服务器和浏览器中，从而保证用户都可以与 Web 站点安全通信。具体要求如下。

（1）在 Master 上，配置 DNS 服务器，设置 IP 地址为 172.16.1.22/24，域名为 "phei.com.cn"。

（2）创建基于域名的虚拟主机，配置 Apache 虚拟主机如表 38.1 所示。

表 38.1　配置 Apache 虚拟主机

项　　目	说　　明
基于域名的虚拟主机	IP 地址：172.16.1.22/24 DNS 地址：172.16.1.22 域名：www.phei.com.cn 根目录：/vh/apache 首页内容："This is apache homepage." 首页文件：index.html

（3）实现 Apache 服务器和 Nginx 服务器的 SSL 访问。

（4）在 Client 上，使用"curl"命令测试 Web 服务器安全访问配置的正确性。

任务实施

步骤 1：配置 Apache 服务器的 SSL 访问。

（1）检验 OpenSSL 软件包是否安装成功，如下所示。

```
root@master:~# dpkg -l openssl
Desired=Unknown/Install/Remove/Purge/Hold
| Status=Not/Inst/Conf-files/Unpacked/halF-conf/Half-inst/trig-aWait/Trig-pend
|/ Err?=(none)/Reinst-required (Status,Err: uppercase=bad)
||/ Name           Version            Architecture Description
+++-==============-==================-============-================================
ii  openssl        3.0.2-0ubuntu1.12  amd64        Secure Sockets Layer toolkit - cryptographic utility
//ii 表示已安装成功
```

（2）安装 Apache 软件包。使用"apt install -y apache2"命令安装 Apache 服务器需要的软件包，如下所示。

```
root@master:~# apt install -y apache2
```

（3）使用"a2enmod"命令启用支持 SSL 加密的 Apache 模块，如下所示。

```
root@master:~# a2enmod ssl
Considering dependency setenvif for ssl:
Module setenvif already enabled
Considering dependency mime for ssl:
Module mime already enabled
Considering dependency socache_shmcb for ssl:
Enabling module socache_shmcb.
Enabling module ssl.
See /usr/share/doc/apache2/README.Debian.gz on how to configure SSL and create self-signed certificates.
To activate the new configuration, you need to run:
  systemctl restart apache2
```

（4）创建私钥。使用"openssl"命令在/etc/apache2/ssl/目录下创建一个 2048 位的私钥 server.key，如下所示。

```
root@master:~# mkdir /etc/apache2/ssl
root@master:~# openssl genrsa -out /etc/apache2/ssl/server.key 2048
```

（5）生成数字证书签署请求。通过新创建的私钥生成一个数字证书签署请求，如下所示。

```
root@master:~# openssl req -new -key /etc/apache2/ssl/server.key -out /etc/apache2/ssl/server.csr
```

```
You are about to be asked to enter information that will be incorporated
into your certificate request.
What you are about to enter is what is called a Distinguished Name or a DN.
There are quite a few fields but you can leave some blank
For some fields there will be a default value,
If you enter '.', the field will be left blank.
-----
Country Name (2 letter code) [AU]:CN                              //国家代码
State or Province Name (full name) [Some-State]:Guangdong          //州或省的全称
Locality Name (eg, city) []:Zhuhai                                //地区
Organization Name (eg, company) [Internet Widgits Pty Ltd]:phei   //公司
Organizational Unit Name (eg, section) []:phei                    //部门
Common Name (e.g. server FQDN or YOUR name) []:phei.com.cn        //域名
Email Address []:                                                 //邮箱地址

Please enter the following 'extra' attributes
to be sent with your certificate request
A challenge password []:                                          //直接按"Enter"键
An optional company name []:                                      //直接按"Enter"键
```

（6）生成自签名数字证书。使用"openssl"命令创建自签名数字证书，如下所示。

```
root@master:~# openssl x509 -req -days 365 -in /etc/apache2/ssl/server.csr -signkey /etc/apache2/ssl/server.key -out /etc/apache2/ssl/server.crt
Certificate request self-signature ok
subject=C = CN, ST = Guangdong, L = Zhuhai, O = phei, CN = phei.com.cn
```

> **小贴士**
>
> 由于自签名数字证书没有经过第三方认证，自签名数字证书一般不可以用于生成环境，仅可以用于开发或测试。

（7）配置 Apache 服务器使用自签名数字证书。生成/etc/apache2/sites-available/vhost1-ssl.conf 文件，并配置 Apache 虚拟主机、管理员邮箱、指定文档根目录和自签名数字证书，如下所示。

```
root@master:~# cd /etc/apache2/sites-available/
root@master:/etc/apache2/sites-available# cp default-ssl.conf vhost1-ssl.conf
root@master:/etc/apache2/sites-available# vim vhost1-ssl.conf
    1 <IfModule mod_ssl.c>
    2     <VirtualHost *:443>
    3         ServerAdmin webmaster@phei.com.cn
    4         ServerName www.phei.com.cn
    5         DocumentRoot /vh/www
    6         <Directory /vh/www1>
```

```
    7              AllowOverride none
    8              Require all granted
    9          </Directory>
……                                      //此处省略部分内容
   29          SSLEngine on
……                                      //此处省略部分内容
   35          #   SSLCertificateFile directive is needed.
   36          SSLCertificateFile      /etc/apache2/ssl/server.crt
   37          SSLCertificateKeyFile /etc/apache2/ssl/server.key
……                                      //此处省略部分内容
  134      </VirtualHost>
  135 </IfModule>
```

（8）为网站创建文档根目录和首页文件，如下所示。

```
root@master:~# mkdir -p /vh/www
root@master:~# echo "This is Apache ssl homepage.">/vh/www/index.html
```

（9）只有使用"a2ensite"命令激活 vhost1-ssl.conf 文件中的配置，服务器的站点内容才能正常显示，如下所示。

```
root@master:/etc/apache2/sites-available# a2ensite vhost1-ssl.conf
Enabling site vhost1-ssl.
To activate the new configuration, you need to run:
  systemctl reload apache2
```

（10）重启 Apache 服务，并设置开机自动启动，如下所示。

```
root@master:~# systemctl restart apache2
root@master:~# systemctl enable apache2
```

步骤2：配置 Nginx 服务器的 SSL 访问。

（1）检验 OpenSSL 软件包是否安装成功，如下所示。

```
root@master:~# dpkg -l openssl
```

（2）安装 Nginx 软件包。使用"apt install -y nginx"命令安装 Apache 服务器需要的软件包，如下所示。

```
root@master:~# apt install -y nginx
```

（3）创建私钥。使用"openssl"命令在/etc/nginx/ssl/目录下创建一个 2048 位的私钥 server.key，如下所示。

```
root@master:~# mkdir /etc/nginx/ssl
root@master:~# openssl genrsa -out /etc/nginx/ssl/server.key 2048
```

（4）生成数字证书签署请求。通过新创建的私钥生成一个数字证书签署请求，如下所示。

```
root@master:~# openssl req -new -key /etc/nginx/ssl/server.key -out /etc/nginx/ssl/server.csr
You are about to be asked to enter information that will be incorporated
```

```
into your certificate request.
What you are about to enter is what is called a Distinguished Name or a DN.
There are quite a few fields but you can leave some blank
For some fields there will be a default value,
If you enter '.', the field will be left blank.
-----
Country Name (2 letter code) [AU]:CN                                //国家代码
State or Province Name (full name) [Some-State]:Guangdong           //州或省的全称
Locality Name (eg, city) []:Zhuhai                                  //地区
Organization Name (eg, company) [Internet Widgits Pty Ltd]:phei     //公司
Organizational Unit Name (eg, section) []:phei                      //部门
Common Name (e.g. server FQDN or YOUR name) []:phei.com.cn          //域名
Email Address []:                                                   //邮箱地址

Please enter the following 'extra' attributes
to be sent with your certificate request
A challenge password []:                                            //直接按"Enter"键
An optional company name []:                                        //直接按"Enter"键
```

（5）生成自签名数字证书。使用"openssl"命令创建自签名数字证书，如下所示。

```
root@master:~# openssl x509 -req -days 365 -in /etc/nginx/ssl/server.csr -signkey /etc/nginx/ssl/server.key -out /etc/nginx/ssl/server.crt
Certificate request self-signature ok
subject=C = CN, ST = Guangdong, L = Zhuhai, O = phei, CN = phei.com.cn
```

（6）配置 Nginx 服务器使用自签名数字证书。生成/etc/nginx/sites-available/ssl 文件，并配置 Nginx 虚拟主机、管理员邮箱、指定文档根目录和自签名数字证书，如下所示。

```
root@master:~# cd /etc/nginx/sites-available/
root@master:/etc/nginx/sites-available# cp default ssl
root@master:/etc/nginx/sites-available# vim ssl
    //在文件的最后添加如下内容
    server {
            listen 443 ssl;
            server_name www.phei.com.cnm;
            root /vh/www/;
            index index.html;
            ssl_certificate /etc/nginx/ssl/server.crt;
            ssl_certificate_key /etc/nginx/ssl/server.key;
            location / {
                    try_files $uri $uri/ =404;
            }
    }
```

（7）为网站创建文档根目录和首页文件，如下所示。

```
root@master:~# mkdir -p /vh/www
```

```
root@master:~# echo "This is Nginx ssl homepage.">/vh/www/index.html
```

（8）重启 Nginx 服务，并设置开机自动启动，如下所示。

```
root@master:~# systemctl restart nginx
root@master:~# systemctl enable nginx
```

任务验收

（1）在 Client 中，查看 Client 的 DNS 地址指向的服务器，确保两台主机之间的网络连接正常。

```
root@client:~# cat /etc/resolv.conf        //查看 Client 的 DNS 地址指向的服务器
nameserver 172.16.1.22
```

（2）在 Client 的命令提示符窗口中，使用 "curl" 命令对 Apache 服务器的 SSL 访问进行测试，如下所示。

```
root@client:~# curl -k https://www.phei.com.cn
This is Apache ssl homepage.
```

（3）在 Master 中，停止 Apache 服务，如下所示。

```
root@master:~# systemctl stop apache2
```

（4）在 Client 的命令提示符窗口中，使用 "curl" 命令对 Nginx 服务器的 SSL 访问进行测试，如下所示。

```
root@client:~# curl -k https://www.phei.com.cn
This is Nginx ssl homepage.
```

任务 39　配置 FTP 服务器

任务目标

1．了解 FTP 服务器的工作过程。
2．熟悉 vsftpd 的配置文件。
3．合理分配不同账户的权限。

任务描述

某公司架设的 Web 服务器需要经常在网站上上传和更新资料，因此需要通过架设 FTP 服务器来完成文件上传和下载。现在公司网络管理员小赵需要安装 FTP 软件包，并将其配置为 FTP 服务器。

任务要求

FTP 服务器的配置主要是通过修改 vsftpd 服务的配置文件来实现的，然而这些配置对 Linux 的初学者而言是比较困难的，因此小赵请来公司的工程师帮助完成 FTP 服务器的配置。配置 FTP 服务器的网络拓扑结构，如图 39.1 所示。

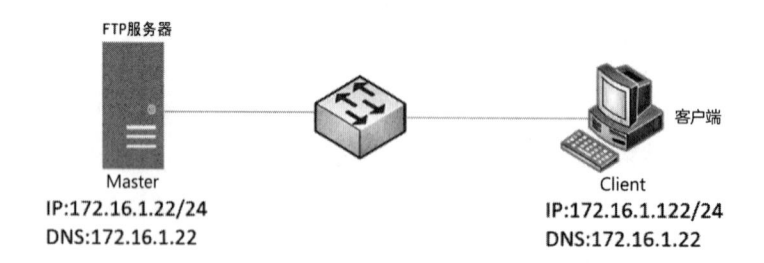

图 39.1　配置 FTP 服务器的网络拓扑结构

具体要求如下。
（1）设置公司 FTP 服务器的 IP 地址为 172.16.1.22/24。
（2）安装 FTP 软件包，并启用匿名用户登录，匿名用户对/srv/ftp/phei 目录具有上传、下载、创建目录和文件、删除目录和文件等权限。

（3）配置实名用户具有上传、下载等权限，其中 ftpuser1 用户被限制在自己的主目录下操作，ftproot 用户可以向上切换目录操作。

任务实施

步骤 1：配置 Master 的 IP 地址等信息，在前面的任务中已经介绍，这里不再详述。

步骤 2：安装 FTP 软件包。

使用"apt install -y vsftpd"命令安装 FTP 服务器需要的软件包，如下所示。

```
root@master:~# apt install -y vsftpd
```

步骤 3：创建/srv/ftp/phei 目录，开放其他用户完全控制的权限，如下所示。

```
root@master:~# mkdir /srv/ftp/phei
root@master:~# chmod 777 /srv/ftp/phei
root@master:~# cd /srv/ftp/phei
root@master:~# mkdir zzb                    //用于测试的文件夹
root@master:~# touch wq.txt                 //用于测试的文件
root@master:~# chmod o=rwx wq.txt           //赋予其他用户完全控制的权限
```

步骤 4：添加本地用户 ftpuser1 和 ftproot，并分别设置登录密码，如下所示。

```
root@master:~# useradd -m ftpuser1
root@master:~# useradd -m ftproot
root@master:~# passwd ftpuser1
root@master:~# passwd ftproot
```

步骤 5：配置 FTP 服务器。

（1）根据任务要求，在/etc/vsftpd.conf 配置文件中修改以下关于匿名用户的配置项。

```
root@master:~# vim /etc/vsftpd.conf
anon_enable=YES
write_enable=YES
anon_upload_enable=YES
anon_umask=022
anon_mkdir_write_enable=YES
anon_other_write_enable=YES
no_anon_passwd=YES
```

（2）根据任务要求，在/etc/vsftpd.conf 配置文件中修改以下关于本地用户的配置项。

```
root@master:~# vim /etc/vsftpd.conf
local_enable=YES
write_enable=YES
local_umask=022
chroot_local_user=YES
```

```
chroot_list_enable=YES
chroot_list_file=/etc/vsftpd.chroot.list
root@master:~# echo "ftproot" > /etc/vsftpd.chroot.list
```

知识链接：

/etc/vsftpd.conf 配置文件都是以"选项名=选项值"的形式定义的，其常用选项及功能如表 39.1 所示。

表 39.1 /etc/vsftpd.conf 配置文件中的常用选项及功能

选 项	默 认 值	功 能
anonymous_enable	NO	是否允许匿名用户登录 vsftpd
anon_upload_enable	NO	是否允许匿名用户上传文件。如果允许，则配置全局 write_enable=YES
anon_mkdir_write_enable	NO	是否允许匿名用户在一定条件下创建目录
anon_other_write_enable	NO	是否允许匿名用户其他的写入操作，如删除和重命名等
anon_umask	077	匿名用户创建文件的权限掩码
anon_root	/srv/ftp	匿名用户登录成功后的默认路径
anon_max_rate	0	匿名用户的最大传输速度，单位为秒，0 表示无限制
no_anon_password	NO	是否询问匿名用户的密码
local_enable	NO	是否允许本地用户登录
write_enable	NO	是否允许任何形式的 FTP 写入命令
local_umask	077	本地用户创建文件的权限掩码
chroot_local_user	NO	是否将本地用户限制在主目录中
chroot_list_enable	NO	是否启用 chroot_list_file 配置项指定的用户列表
chroot_list_file	/etc/vsftpd.chroot.list	指定被限制在主目录中的用户列表
dirmessage_enable	NO	指定用户首次进入新目录时是否显示消息
xferlog_enable	NO	是否允许上传/下载日志记录
connect_from_port_20	NO	是否使用 20 号端口来传输数据
chown_upload	NO	是否将匿名用户上传的文件所有者更改为 chown_username 选项指定的用户
chown_username	root	匿名用户上传文件的默认所有者
xferlog_file	/var/log/xferlog	设置日志文件的保存位置

步骤 6：在配置完成后，重启 FTP 服务，并设置开机自动启动，如下所示。

```
root@master:~# systemctl restart vsftpd
root@master:~# systemctl enable vsftpd
```

任务验收

（1）在 Linux 客户端进行匿名用户测试。

① 设置 Linux 客户端和 Samba 服务器之间的网络连通。

② 先使用匿名用户登录 FTP 服务器并进入 phei 目录，再使用"get"交互命令将 upload.txt 文件下载到本地，将 download.txt 文件上传到 FTP 服务器的 phei 目录，并修改 ML 目录的名称为"ZW"，测试如下所示。

```
root@ubuntu:~# ftp 192.168.1.201
Connected to 192.168.1.201.
220 (vsFTPd 3.0.5)
Name (192.168.1.201:root): anonymous
230 Login successful.
Remote system type is UNIX.
Using binary mode to transfer files.
ftp> !pwd                               //查看本地目录
/root
ftp> !ls                                //查看本地目录的内容
snap
ftp> ls                                 //查看服务器目录
229 Entering Extended Passive Mode (|||65044|)
150 Here comes the directory listing.
drwxrwxrwx    3 0        0            4096 Jan 10 15:10 phei
226 Directory send OK.
ftp> cd phei
250 Directory successfully changed.
ftp> ls
229 Entering Extended Passive Mode (|||16226|)
150 Here comes the directory listing.
drwxr-xr-x    2 115      121          4096 Jan 10 15:08 ML
-rw-r--r--    1 115      121             0 Jan 10 15:08 upload.txt
226 Directory send OK.
ftp> get upload.txt                     //下载 upload.txt 文件
local: upload.txt remote: upload.txt
229 Entering Extended Passive Mode (|||31110|)
150 Opening BINARY mode data connection for upload.txt (0 bytes).
0       0.00 KiB/s
226 Transfer complete.
ftp> !ls                                //查看本地文件
download.txt  snap  upload.txt
ftp> put download.txt                   //上传 download.txt 文件
local: download.txt remote: download.txt
```

```
229 Entering Extended Passive Mode (|||30142|)
150 Ok to send data.
    0        0.00 KiB/s
226 Transfer complete.
ftp> ls
229 Entering Extended Passive Mode (|||19533|)
150 Here comes the directory listing.
drwxr-xr-x    2 115      121          4096 Jan 11 12:37 ML
-rw-r--r--    1 115      121             0 Jan 11 13:17 download.txt
-rw-r--r--    1 115      121             0 Jan 11 12:37 upload.txt
226 Directory send OK.
ftp> rename ML ZW                              //重命名
350 Ready for RNTO.
250 Rename successful.
ftp> ls                                        //查看是否生效
229 Entering Extended Passive Mode (|||43491|)
150 Here comes the directory listing.
drwxr-xr-x    2 115      121          4096 Jan 11 12:37 ZW
-rw-r--r--    1 115      121             0 Jan 11 13:17 download.txt
-rw-r--r--    1 115      121             0 Jan 11 12:37 upload.txt
226 Directory send OK.
ftp> bye                                       //退出 FTP 客户端
221 Goodbye.
```

（2）在 Linux 客户端进行本地用户测试。

① 设置 Linux 客户端和 FTP 服务器之间的网络连通。

② 使用 ftpuser1 用户身份登录 FTP 服务器，查看当前目录为"/"，并创建目录测试写权限，向上切换目录后使用"pwd"命令查看当前目录，结果仍为"/"，表明 ftpuser1 被限定在用户的主目录下，测试如下所示。

```
root@ubuntu:~# ftp 192.168.1.201
Connected to 192.168.1.201.
220 (vsFTPd 3.0.5)
Name (192.168.1.201:root): ftpuser1
331 Please specify the password.
Password:
230 Login successful.
Remote system type is UNIX.
Using binary mode to transfer files.
ftp> pwd
Remote directory: /
ftp> ls
229 Entering Extended Passive Mode (|||27297|)
150 Here comes the directory listing.
```

```
drwxr-xr-x    2 1001     1001         4096 Jan 07 14:36 ftpuser1dir
226 Directory send OK.
ftp> cd ..
250 Directory successfully changed.
ftp> pwd
Remote directory: /
```

③ 使用 ftproot 用户身份登录 FTP 服务器，查看当前目录显示为"/home/ftproot"，并创建目录测试写权限，使用"cd .."命令向上切换目录，可以看到结果为"/home"，表明 ftproot 用户可以切换到用户主目录外，测试结果如下所示。

```
root@ubuntu:~# ftp 192.168.1.201
Connected to 192.168.1.201.
220 (vsFTPd 3.0.5)
Name (192.168.1.201:root): ftproot
331 Please specify the password.
Password:
230 Login successful.
Remote system type is UNIX.
Using binary mode to transfer files.
ftp> pwd
Remote directory: /home/ftproot
ftp> ls
229 Entering Extended Passive Mode (|||49971|)
150 Here comes the directory listing.
drwxr-xr-x    2 1002     1002         4096 Jan 07 14:37 root-dir
226 Directory send OK.
ftp> mkdir ftproot-dir
257 "/home/ftproot/ftproot-dir" created
ftp> ls
229 Entering Extended Passive Mode (|||12783|)
150 Here comes the directory listing.
drwxr-xr-x    2 1002     1002         4096 Jan 10 15:26 ftproot-dir
drwxr-xr-x    2 1002     1002         4096 Jan 07 14:37 root-dir
226 Directory send OK.
ftp> cd ..
250 Directory successfully changed.
ftp> pwd
Remote directory: /home
```

任务 40　配置 MySQL 数据库服务器

任务目标

1. 掌握 MySQL 数据库服务器的安装和配置。
2. 使用命令实现数据库和数据表的基本操作。

任务描述

某公司的网络管理员小赵根据公司的业务需求，需要在信息中心的 Linux 服务器上搭建数据库，以满足公司搭建 OA 办公系统的需求。小赵需要安装 MySQL 数据库软件包，并将其配置为 MySQL 数据库服务器。

任务要求

为满足公司的需求，需要先安装 MySQL 数据库服务器，再对 MySQL 数据库服务器进行配置，包括数据库的创建、数据表的创建和对数据表增、删、改、查等。MySQL 数据库服务器的配置主要是通过命令的操作实现数据库的功能。具体要求如下。

（1）安装 MySQL 数据库软件包，将其配置为 MySQL 数据库服务器。

（2）创建 myschool 数据库，并在其中创建 mystudent 数据表。

（3）在数据表中创建 2 个用户，分别为 user1（序号为 202308001，用户名为 myuser1，出生日期为 1996-7-1，性别为 male）、user2（序号为 202308002，用户名为 myuser2，出生日期为 1996-9-1，性别为 female），密码与用户名相同，mystudent 数据表结构如表 40.1 所示。

表 40.1　mystudent 数据表结构

字 段 名 称	数 据 类 型	主　　键
ID	Int	是
Name	varchar(10)	否

字 段 名 称	数 据 类 型	主　　键
Birthday	Datetime	否
Sex	char(8)	否
Password	char（128）	否

（4）创建名为"teacher"的仅可本地登录的用户，并设置该用户具备所有权限，密码自行设置。

（5）对 myschool 数据库进行备份，并设置在发生故障时和 myschool 数据库被损坏时可以恢复。

任务实施

步骤 1：安装 MySQL 数据库软件包，如下所示。

```
root@ubuntu:~# apt install -y mysql-server
root@ubuntu:~# dpkg -l mysql-server
Desired=Unknown/Install/Remove/Purge/Hold
| Status=Not/Inst/Conf-files/Unpacked/halF-conf/Half-inst/trig-aWait/Trig-pend
|/ Err?=(none)/Reinst-required (Status,Err: uppercase=bad)
||/ Name              Version                    Architecture Description
+++-=================-==========================-============-====================
ii  mysql-server      8.0.35-0ubuntu0.22.04.1    all          MySQL database server
(metapackage depending on the latest version)
/ii 表示已安装成功
```

步骤 2：创建数据库并将其命名为"myschool"，在 myschool 数据库中创建 mystudent 数据表，命令如下所示。

```
root@ubuntu:~# mysql -u root -p              //密码默认为空，直接按"Enter"键即可
Enter password:

mysql> create database myschool;             //创建 myschool 数据库
Query OK, 1 row affected (0.01 sec)
mysql> show databases;
+--------------------+
| Database           |
+--------------------+
| information_schema |
| myschool           |
| mysql              |
| performance_schema |
| sys                |
+--------------------+
```

```
5 rows in set (0.01 sec)
mysql> use myschool;
mysql> create table mystudent(ID int primary key,Name varchar(10),Birthday
Datetime,Sex char(8),Password char(128));            //创建 mystudent 数据表
```

知识链接：

MySQL 数据库和数据表常用的命令及功能如表 40.2 所示。

表 40.2　数据库和数据表常用的命令及功能

命　　令	功　　能
show databases	显示当前已有的数据库
show table	显示当前数据库中的数据表
create　database 数据库名称	创建新的数据库
create table 数据表名称（字段名称 字段类型 字段长度……）	创建新的数据表
drop database 数据库名称	删除数据库
drop table 数据表名称	删除数据表
use 数据库名称	切换数据库
Desc 数据表名称	显示数据表的结构
insert into 数据表名称 values ('数据1'……)	向数据表中录入一条记录
select * from 数据表名称	查看数据表中的所有记录（*表示所有）
delete from 数据表名称 where 字段名称=值	删除符合条件的记录

步骤 3：在数据表中创建 2 个用户，密码与用户名相同，命令如下所示。

```
mysql> insert into mystudent values(202308001,'myuser1','1996-7-1','male','myuser1');
Query OK, 1 row affected (0.00 sec)
mysql> insert into mystudent values(202308002,'myuser2','1996-9-1','female','myuser2');
Query OK, 1 row affected (0.00 sec)
```

步骤 4：创建名为"teacher"的仅可本地登录的用户，让该用户具备所有权限，密码自行设置，命令如下所示。

```
mysql> create user 'teacher'@'localhost' identified by '123456';
Query OK, 0 rows affected (0.01 sec)
mysql> grant all privileges on *.* to 'teacher'@'localhost' with grant option;
Query OK, 0 rows affected (0.01 sec)
```

步骤 5：对 myschool 数据库进行备份。

（1）使用"mysqldump"命令将数据库导出到指定目录下并保存，查看备份文件，如下所示（在数据库备份前，mystudent 数据表中已有 2 条记录）。

```
root@ubuntu:~# mkdir mysqlbak
```

```
root@ubuntu:~# mysqldump myschool -u root -p > /root/mysqlbak/myschool_bak.sql
Enter password:
```

（2）删除数据库。使用"drop database"命令彻底删除 myschool 数据库，并显示当前所有数据库，如下所示。

```
mysql> show databases;                      //查询数据库
+--------------------+
| Database           |
+--------------------+
| information_schema |
| mysql              |
| performance_schema |
| myschool           |
+--------------------+
4 rows in set (0.01 sec)
mysql> drop database myschool;              //删除数据库
Query OK, 1 row affected (0.02 sec)
mysql> show databases;                      //查询数据库
+--------------------+
| Database           |
+--------------------+
| information_schema |
| mysql              |
| performance_schema |
| sys                |
+--------------------+
4 rows in set (0.00 sec)
//myschool 数据库已被删除
```

步骤 6：恢复 myschool 数据库。

（1）使用命令登录 MySQL 数据库后，创建空数据库 myschool，并查看数据库中的数据表，如下所示。

```
mysql> create database myschool;
Query OK, 1 row affected (0.00 sec)
mysql> use myschool;
Database changed
mysql> show tables;                         //查看数据库内的表格
Empty set (0.00 sec)
```

（2）使用重定向符"<"将备份的数据库文件导入刚创建的空数据库中，如下所示。

```
root@ubuntu:~# mysql -u root -p myschool < /root/mysqlbak/myschool_bak.sql
Enter password:
root@ubuntu:~# mysql -u root -p
Enter password:
```

```
Welcome to the MySQL monitor.  Commands end with ; or \g.
Your MySQL connection id is 12
Server version: 8.0.35-0ubuntu0.22.04.1 (Ubuntu)

Copyright (c) 2000, 2023, Oracle and/or its affiliates.

Oracle is a registered trademark of Oracle Corporation and/or its
affiliates. Other names may be trademarks of their respective
owners.

Type 'help;' or '\h' for help. Type '\c' to clear the current input statement.
mysql> use myschool;
Database changed
mysql> show tables;
+--------------------+
| Tables_in_myschool |
+--------------------+
| mystudent          |
+--------------------+
1 row in set (0.00 sec)
```

知识链接：

在MySQL中，每个数据库和数据表分别对应文件系统中的目录和其下的文件。在Linux操作系统中，数据库文件的存放目录一般为/var/lib/mysql。

（1）备份数据库。

"mysqldump"命令用于备份数据库，基本语法格式如下。

```
mysqldump -user=root -password=root密码 数据库名 > 备份文件.sql
```

（2）恢复数据库。

在恢复数据库时，需要先创建好一个数据库（不一定同名），再将备份出来的文件导入新创建的数据库中。

"mysql"命令用于恢复数据库，基本语法格式如下。

```
mysql -u root -password=root密码 数据库名 < 备份文件.sql
```

（3）使用命令查看导入的数据库中的数据表结构和记录，如下所示。

```
mysql> desc mystudent;                        //查看数据表结构
+-----------+-------------+------+-----+---------+-------+
| Field     | Type        | Null | Key | Default | Extra |
+-----------+-------------+------+-----+---------+-------+
| ID        | int         | NO   | PRI | NULL    |       |
| Name      | varchar(10) | YES  |     | NULL    |       |
```

```
| Birthday     | datetime    | YES   |       | NULL      |       |
| Sex          | char(8)     | YES   |       | NULL      |       |
| Password     | char(128)   | YES   |       | NULL      |       |
+--------------+-------------+-------+-------+-----------+-------+
5 rows in set (0.00 sec)
mysql> select * from mystudent;
+-----------+---------+---------------------+--------+----------+
| ID        | Name    | Birthday            | Sex    | Password |
+-----------+---------+---------------------+--------+----------+
| 202308001 | myuser1 | 1996-07-01 00:00:00 | male   | myuser1  |
| 202308002 | myuser2 | 1996-09-01 00:00:00 | female | myuser2  |
+-----------+---------+---------------------+--------+----------+
2 rows in set (0.00 sec)
```

任务验收

（1）使用"desc"命令查看 mystudent 数据表的结构，命令如下所示。

```
mysql> desc mystudent;                              //查看数据表结构
+--------------+-------------+-------+-------+---------+-------+
| Field        | Type        | Null  | Key   | Default | Extra |
+--------------+-------------+-------+-------+---------+-------+
| ID           | int         | NO    | PRI   | NULL    |       |
| Name         | varchar(10) | YES   |       | NULL    |       |
| Birthday     | datetime    | YES   |       | NULL    |       |
| Sex          | char(8)     | YES   |       | NULL    |       |
| Password     | char(128)   | YES   |       | NULL    |       |
+--------------+-------------+-------+-------+---------+-------+
5 rows in set (0.02 sec)
```

（2）使用"select"命令查看 mystudent 数据表中的内容，命令如下所示。

```
mysql> select * from mystudent;
+-----------+---------+---------------------+--------+----------+
| ID        | Name    | Birthday            | Sex    | Password |
+-----------+---------+---------------------+--------+----------+
| 202308001 | myuser1 | 1996-07-01 00:00:00 | male   | myuser1  |
| 202308002 | myuser2 | 1996-09-01 00:00:00 | female | myuser2  |
+-----------+---------+---------------------+--------+----------+
2 rows in set (0.00 sec)
```

（3）使用"select"命令查看 teacher 用户是否创建成功，命令如下所示。

```
mysql> select user from user;
+------------------+
| user             |
+------------------+
```

```
| debian-sys-maint  |
| mysql.infoschema  |
| mysql.session     |
| mysql.sys         |
| root              |
| teacher           |                        //teacher 用户已存在
+-------------------+
6 rows in set (0.00 sec)
```

（4）使用"show grants"命令查看 teacher 用户的权限，命令如下所示。

```
mysql> show grants for 'teacher'@'localhost' ;
……                                         //省略部分内容
 | GRANT SELECT, INSERT, UPDATE, DELETE, CREATE, DROP, RELOAD, SHUTDOWN, PROCESS,
FILE, REFERENCES, INDEX, ALTER, SHOW DATABASES, SUPER, CREATE TEMPORARY TABLES, LOCK
TABLES, EXECUTE, REPLICATION SLAVE, REPLICATION CLIENT, CREATE VIEW, SHOW VIEW, CREATE
ROUTINE, ALTER ROUTINE, CREATE USER, EVENT, TRIGGER, CREATE TABLESPACE, CREATE ROLE,
DROP ROLE ON *.* TO `teacher`@`localhost` WITH GRANT OPTION              |
 | GRANT
APPLICATION_PASSWORD_ADMIN,AUDIT_ABORT_EXEMPT,AUDIT_ADMIN,AUTHENTICATION_POLICY_ADMIN,BA
CKUP_ADMIN,BINLOG_ADMIN,BINLOG_ENCRYPTION_ADMIN,CLONE_ADMIN,CONNECTION_ADMIN,ENCRYPTION_
KEY_ADMIN,FIREWALL_EXEMPT,FLUSH_OPTIMIZER_COSTS,FLUSH_STATUS,FLUSH_TABLES,FLUSH_USER_RES
OURCES,GROUP_REPLICATION_ADMIN,GROUP_REPLICATION_STREAM,INNODB_REDO_LOG_ARCHIVE,INNODB_R
EDO_LOG_ENABLE,PASSWORDLESS_USER_ADMIN,PERSIST_RO_VARIABLES_ADMIN,REPLICATION_APPLIER,RE
PLICATION_SLAVE_ADMIN,RESOURCE_GROUP_ADMIN,RESOURCE_GROUP_USER,ROLE_ADMIN,SENSITIVE_VARI
ABLES_OBSERVER,SERVICE_CONNECTION_ADMIN,SESSION_VARIABLES_ADMIN,SET_USER_ID,SHOW_ROUTINE
,SYSTEM_USER,SYSTEM_VARIABLES_ADMIN,TABLE_ENCRYPTION_ADMIN,TELEMETRY_LOG_ADMIN,XA_RECOVE
R_ADMIN ON *.* TO `teacher`@`localhost` WITH GRANT OPTION |
   +------------------------------------------------------------------------
……                                         //省略部分内容
2 rows in set (0.00 sec)
```

反侵权盗版声明

电子工业出版社依法对本作品享有专有出版权。任何未经权利人书面许可，复制、销售或通过信息网络传播本作品的行为；歪曲、篡改、剽窃本作品的行为，均违反《中华人民共和国著作权法》，其行为人应承担相应的民事责任和行政责任，构成犯罪的，将被依法追究刑事责任。

为了维护市场秩序，保护权利人的合法权益，我社将依法查处和打击侵权盗版的单位和个人。欢迎社会各界人士积极举报侵权盗版行为，本社将奖励举报有功人员，并保证举报人的信息不被泄露。

举报电话：（010）88254396；（010）88258888
传　　真：（010）88254397
E-mail: dbqq@phei.com.cn
通信地址：北京市万寿路 173 信箱
　　　　　电子工业出版社总编办公室
邮　　编：100036